Urban Movements and Climate Change

Urban Movements and Climate Change

Loss, Damage and Radical Adaptation

Edited by
Marco Armiero,
Ethemcan Turhan and
Salvatore Paolo De Rosa

Routledge
Taylor & Francis Group
LONDON AND NEW YORK

The publication of this book is made possible thanks to a grant from FORMAS (Swedish Research Council, Project No: 2017-01962_3) and KTH Library.

First published in 2023 by Amsterdam University Press Ltd.

Published 2025 by Routledge
4 Park Square, Milton Park, Abingdon, Oxon OX14 4RN
605 Third Avenue, New York, NY 10158

Routledge is an imprint of the Taylor & Francis Group, an informa business

ISBN: 9789463726665 (hbk)
ISBN: 9781041190110 (pbk)
ISBN: 9781003708933 (ebk)
NUR 907

Cover photo: Kulturkvarter Snösätra (www.snosatra.com), Stockholm. Photo by Marco Armiero

Cover design: Coördesign, Leiden

DOI: 10.5117/9789463726665

Table of Contents

List of figures and tables

Figures

Tables

Acknowledgements

This volume would not have been possible were it not for the patience of our authors, our editors at AUP and the collegial support and comradely help of multiple other people. We would particularly like to thank Diego Andreucci, Gustavo García López, Eda Acara, Mar Grau, Sinan Erensü, Ilenia Iengo, Ekin Kurtiç, Jevgeniy Bluwstein, Jesse D. Peterson and Daniele Valisena for their invaluable feedback on individual chapters. We gratefully acknowledge the Occupy Climate Change! project funded by FORMAS (Swedish Research Council for Sustainable Development) under the National Research Programme on Climate (Contract: 2017-01962_3).

Acknowledgements

[illegible]

Foreword: From Occupy Climate Change! to confronting loss and damage

David Naguib Pellow

Abstract
David N. Pellow places this volume within the scholarship on climate and environmental justice. Building upon the chapters gathered in this volume, Pellow maintains that the climate crisis should not be reduced to a problem of carbon emissions but framed in terms of a radical transformation of every aspect of society. Pellow, along with the authors of this volume, argues for a repoliticization of the climate crisis, reminding us that the concept of loss and damage, now employed even in international negotiations, is the result of grassroots struggles pointing to the unjust distribution of the effects of climate change. Embracing the urban focus of the volume, Pellow evokes a progressive—perhaps revolutionary—politics that is able to rebuild and repurpose, hack and reimagine the city in ways that facilitate equity, justice, sustainability and regeneration within and beyond the urban centers.

Keywords: climate justice, loss and damage, repoliticization of climate change, repairing the world

This extraordinary book has been many years in the making, from its inception as a project conceived under the title *Occupy Climate Change!* some years ago, through an invigorating workshop in Stockholm in 2020. Many upheavals and major world events have shaped this initiative along the way, delaying its release but offering space and time for additional reflection and adjustment to address the rapidly changing terrain upon which we rest and work. The context for this collective effort includes (but is certainly not limited to) the ongoing catastrophe of global climate change and disruption, the sixth mass extinction of untold populations of

Armiero, M., E. Turhan & S. Paolo De Rosa (eds), *Urban Movements and Climate Change: Loss, Damage and Radical Adaptation*. Taylor & Francis Group 2023
DOI: 10.5117/9789463726665_FORE

flora and fauna, the amplification of racial injustices and state-sanctioned racist violence and the scourge of the COVID-19 pandemic, all of which are deeply intertwined and produce terrifying and inequitable intersections and ripple effects. That's the bad news, of course, but the more uplifting context in which this book emerges is the fact that we are living in a time when the majority of humans are residents of urban centers where millions of people are articulating their demands and enacting practices that reflect the right to the city and the hopeful power of urban climate justice movements that hold the most promising solutions to the crises we face as a global multispecies community. Cities are at the heart of this volume because, as Karen Paiva Henrique notes, they are sites of unsustainable elite consumption, they are also hotspots where we observe and experience major climate events, and because they are "beacons of hope" for innovative practices that just might lead us toward radical pathways to address the climate crisis.

The earlier moment of incubation for this book—the Occupy Climate Change! framing—is important because the term "occupy" is deployed here not in a settler colonial posture, but rather in its much older tradition as a vision and practice of reclaiming and managing spaces and resources through noncapitalist and non-state-dominated methods and means. This idea was inspired by and built on the earlier examples of Occupy Wall Street and Occupy (Hurricane) Sandy, wherein community members worked to provide and meet critical needs and material necessities for people during moments of intensified crisis through mutual aid and a myriad of other creative mechanisms. These movements remain enduring examples of how ordinary people have risen to the occasion to offer essential support for urban dwellers (again, who constitute the majority of our species) when the state and other dominant institutions cannot or will not provide that assistance (what some call "urban fragility"). These are no longer prefigurative movements but are actually existing practices that will need to be in evidence and in motion for the foreseeable future, as climate change-driven disasters and state-centered neglect of working people will only intensify in the coming years. These events are perhaps better thought of not as the core crises, but rather as symptoms of the everyday, underlying and unrelenting violence that constitutes global racial capitalism.

At the time of this writing the 27th Conference of the Parties (COP27) to the United Nations Framework Convention on Climate Change (UNFCCC) just concluded, with a proposal to develop a fund to provide critical "loss and damage" resources for Global South nations. While this proposal is vague in both content and commitment, its existence is undeniably the result of years of climate justice advocacy from below, among community leaders across the world whose ways of life and whose futures have been threatened by climate disruptions perpetrated primarily by wealthy corporations and Global North societies. The concept of loss and damage is meant to reflect the necessity of providing urgently needed support for those communities and nations on the front lines and fence lines of global climate change, places where the people who have contributed the least to creating the problem are suffering the worst from its wide-ranging impacts. These communities have endured the most significant losses and damages associated with global climate change in the form of flooding, temperature rises, harm to agricultural systems and rising morbidity and mortality. And while the UNFCCC's loss and damage proposal is a work in progress at best, it does represent an acknowledgement of the fundamental inequities associated with the ravages of global climate change. The contributors to this volume write persuasively and poetically about the countless activities that residents and grassroots organizations in cities around the world are pursuing to contest and address losses and damages with an eye toward building dynamic and healthy communities that will sustain our multispecies societies for generations to come. Just a sampling of such instructive and uplifting cases where we find encouraging practices and material interventions includes:

- *Public Power NY:* a community-based project that seeks to transition the state of New York's privately run, fossil fuel-centered energy system toward a public, democratically controlled renewable energy initiative (see Dawson's chapter)
- *Social Gardens:* a series of initiatives by activists in the city of Naples, Italy, to reclaim and repurpose lands abandoned by the state for the creation and revitalization of public parks, green spaces, and the production of materials for bodily and psychological sustenance (see Berruti and Palestino's chapter)
- *Reforestation in Babylon Hill:* a collaboration between residents in Brazil's Morro da Babilônia favela and city hall technicians to plant trees to create a revitalized biophysical environment through expanded forest cover, resulting in the realization of both the right to the city and the right to urban nature (see Sedrez and Augosto Barbosa's chapter)
- *Ecobarrios:* an exciting series of bottom-up, self-organized community projects in Santiago, Chile, that are focused on supporting community well-being outside of a market-based system, which includes ecological education, food production, gardening, bartering, time banking, skill shares and the redistribution of surpluses (see Visconti's chapter)

- The development of hip-hop music as an example of a DIY community public works project, born out of a need for working-class communities of color to provide forums for artistic and cultural expression, which resulted in the modification and repairing of city infrastructure to create one of the most defining art forms in all of history (see Khawarzad's chapter)

Each of these cases—and so much more across the book's chapters—underscores a message that cannot be overemphasized: we must not allow the climate crisis to be reduced to a problem of carbon emissions reductions. Instead, we must push to ensure that both the problem and solution are framed around the need to transform and reconstruct every aspect of society and our relationship to our life-supporting ecosystems. In other words, the authors of this volume firmly grasp the need to counter the depoliticization within climate change policy debates with a serious commitment to *re*politicization. That transition requires scholars and activists to confront the "consensual regime of climate change governance," which casts the climate crisis as a "universal problem" that everyone (and therefore no one) is responsible for, and in which only techno-fixes that fit comfortably within the structure of capitalism are allowed. Through counter-hegemonic discourses, tactics and strategies, climate justice activists featured in this book are boldly repoliticizing the climate change question through dissent, disruption and analysis and action that communicates a message that climate futures are chiefly about struggles for power and justice.

The contributors to this volume locate the blame for the global climate crisis squarely at the door of the unrelenting capitalist attack on our commons—the vast trove of social, cultural and ecological wealth that dominant institutions have targeted for privatization, commodification, consumption and despoilation for centuries. Capitalism is a particularly destructive force because it degrades the conditions of its own production, it is premised on infinite expansionary logics and practices and it intensifies the extinction crisis through the creation of a brutal world economic system. I was repeatedly reminded of Ashley Dawson's book *Extreme Cities* (2017), in which he argues clearly and persuasively that these urban areas marked by hyper inequalities are also the same sites that generate large volumes of carbon emissions and that are repeatedly beset by anthropogenic climate change. Contrary to the sunny and, frankly, inaccurate and naïve portrayals of cities as spaces where the beneficence and promises of green capitalism abound, Dawson explains that extreme cities are causing much of our ecological and climate crisis since they require the voracious extraction of ecosystem materials from other sites—a stark portrait of unsustainability. But, contrary to some of the anti-urban and anti-civilizational discourse that has emerged from certain quarters of society and the academy that calls for deurbanization or for an uncritical embrace of the rural, Dawson rightly contends that despite these challenges, cities are also precisely the places where we find workable solutions to our socio-ecological crises. He argues that point not because extreme cities are where there are dense networks of financial institutions and the captains of industry who many pundits believe we should rely on to invent a techno-fix out of climate change, but because these urban centers are the sources of creative and bold grassroots climate and environmental justice movements that recognize that solutions to the climate crisis require addressing deep social and political inequities. The reason why these solutions are the most promising is because Dawson understands a fundamental scientific truth that still has yet to sink in for most of us: social inequality is the single most important and destructive driving force fueling our climate crisis, so those among us who are working to strengthen democracy and equity are focusing on the most significant levers of change required to stabilize and improve humankind's relationship with the planet.

As a scholar who has worked diligently on my own community's version of "environmentalism from below" in general and a Green New Deal, in particular, I find this book's insights to be invaluable because it forces us to ask: How can we transform and decolonize well-meaning and progressive climate policies like the Green New Deal so that they actually speak to the needs and desires of the Global South and other marginalized communities? The contributors offer a range of insightful answers to that question that begin with listening to, learning from and collaborating with leaders and members from those communities. The interventions and analyses from these authors prompted me to reread the "People's Agreement," a public document that emerged from the World People's Conference on Climate Change and the Rights of Mother Nature, held in Cochabamba, Bolivia, in 2010. This document spells out critical lessons and guidance for how to confront the juggernaut of capitalist institutions with a vision and practice of conviviality, care, compassion and resources that can improve the lives and well-being of low wealth people while also strengthening the capacity of critical habitats and ecosystems to regenerate.

The contributors present numerous effective ways of reframing the socio-ecological crises we face as opportunities for deepening our commitments to climate justice and global, multispecies well-being. One of many such examples is Karen Paiva Henrique's use of a feminist theoretical lens that emphasizes centering the experiences of those who exist at the margins of society in order to develop more grounded and transformative understandings of how the climate crisis impacts the world's majorities, as well as the importance of exploring the role of the body in urban climate politics. Panagiota Kotsila proposes that we counter the dominant nativist way of seeing immigrants and climate refugees as a "burden" on our cities to, instead, being seeing them as critical holders of knowledge that can be shared and applied toward addressing the climate crisis. And the volume's editors note the importance of utopian thinking for imagining futures that exceed the limitations of what may be possible in the present. Utopian thinking is perhaps one of our most invaluable and yet underutilized "natural resources."

From despair to repair

My own perspective on these challenging themes is deeply aligned with the revolutionary ideas that the contributors to this collection have articulated. I understand cities as complex spaces where myriad conflicting forces collide and collaborate, offering both dystopian and utopian realities and possibilities. Cities are indeed spaces of mass death and despair, institutional violence, social distancing and deepening inequities. But they are also spaces where climate, environmental, food, social and multispecies justice movements are providing actually existing alternatives to the dominant fossil fuel-driven, neoliberal, eco/genocidal political economy. But I think it is critical to acknowledge that the urban environment must be rebuilt and repurposed, hacked and reimagined in ways that facilitate equity, justice, sustainability and regeneration within the boundaries of the urban centers, but that work must also be done in ways that are equitably linked to the countless rural and far-flung communities whose resources and labor make the city possible—where mining, fishing, deforestation, agriculture and migration occur. Without attention to the rural and nonurban spaces, the city cannot become sustainable or just.

Aurash Khawarzad's contribution to this volume proposes a "user manual" for the city, rooted in the idea of the right to repair the city—an intervention that recognizes that many cities are being built for planned obsolescence and are designed in ways that produce invidious exclusions, precarity and marginalization. The right to repair the city recognizes these forces and seeks to enact practices that create urban spaces marked by equity, radical inclusivity and a healthy quality of life. I would propose that we consider a concept from rabbinic Judaism known as *tikkun olam*, which roughly translates to "repairing the world," as a generative way of extending the right to repair to a global scale. As cities are now the primary places in which the human species resides, repairing the world must necessarily entail repairing cities, and repairing cities will only be possible if we understand how the health of urban centers is intimately linked to the health of the Earth itself. Ultimately, then, the right to the city and the right to repair must also support the multispecies right to a livable and just planet.

References

Dawson, A. (2017). *Extreme cities: The peril and promise of urban life in the age of climate change*. Verso.

"People's Agreement." (2010, 22 April 22). World People's Conference on Climate Change and the Rights of Mother Earth, Cochabamba, Bolivia. https://www.climateemergencyinstitute.com/uploads/Peoples_climate_agreement.pdf

About the author

David Naguib Pellow is the Dehlsen Chair and Distinguished Professor of Environmental Studies and director of the Global Environmental Justice Project at the University of California, Santa Barbara. His research interests include environmental and social justice, race/class/gender and environmental conflict, human-animal conflicts, sustainability and social change movements.

1. Occupy Climate Change! An Introduction

Marco Armiero, Salvatore Paolo De Rosa and Ethemcan Turhan

Abstract

This introduction presents the Occupy Climate Change! research project, the root from which this volume has sprouted. Armiero, De Rosa and Turhan discuss the main themes addressed by the project and the contributors to the volume: the (counter-)power of community led experiments, the trap of the mainstream climate change discourses and policies, and the need to repoliticizing climate adaptation and mitigation. Facing loss and damage now and not in a remote future, communities are experimenting with a wide variety of social innovations, often deeply antagonistic to top-down approaches, sometimes more inclined towards collaborations with institutions. This introduction attempts to systematize the characteristics of social innovations vs. market innovations, though, avoiding to propose any fixed canon to evaluate grassroots experiments.

Keywords: Occupy Climate Change!, social innovations, social movements, urban climate politics, loss and damage

This volume was conceived and finalized from March 2020 to December 2022. In other words, it was squeezed between a global pandemic and a war on the eastern flanks of Europe. It seems that we are transitioning from the COVID-19 pandemic to the risk of nuclear apocalypse, with the climate crisis appearing to go out of fashion in both public discourse and political action. Nonetheless, at least for many Europeans, the ongoing Russo-Ukrainian War and its intersecting crises clearly smell of gas, oil and coal. For the first time in many decades, Europeans were worried about the coming winter. They are realizing that their comfortable lives are still deeply dependent on the availability of relatively cheap fossil fuels. Paradoxically,

Armiero, M., E. Turhan & S. Paolo De Rosa (eds), *Urban Movements and Climate Change: Loss, Damage and Radical Adaptation*. Taylor & Francis Group 2023
DOI: 10.5117/9789463726665_CH01

judging from the absence of a vigorous peace mobilization, it seems that Europeans are more scared of high-priced utilities than they are of nuclear winter and climate breakdown.

Both the pandemic and the war in Ukraine have deeply affected the climate crisis and politics. The lockdown policies implemented by many governments during the pandemic reduced CO_2 emissions in 2020 (Le Quéré et al., 2020), although more recent data indicate that the reduction was strictly temporary (WMO, 2022). The currency of the day is no longer either mitigation or adaptation. It is loss and damage, as the chair of Least Developed Countries group wrote in a recent op-ed. Loss and damage is "irreversible climate-related devastation that cannot be mitigated or adapted to," says Diouf Sarr (2022), who adds, "dealing with it is pushing developing countries into ever-greater debt and their economies to the brink of collapse." The ongoing war is also having direct and indirect effects on climate change. The military operations are responsible for massive CO_2 emissions as well as other harmful effects on the environment, including future impacts that will be caused by reconstruction efforts (Masterson, 2022). In a more systemic sense, the war has caused a dramatic shift in national policies: the imperative has become to reach energy independence from Russia at the cost of resuming the intensive use of coal, LNG (liquefied natural gas) or nuclear power.

When we met in Stockholm at the beginning of March 2020, COVID-19 was already spreading across the planet. We were, obviously, worried about the virus, though at the time we had not fully realized its danger. Just a few days after the workshop one of us got extremely ill with COVID-19. With his life at risk, firsthand experience showed him that the pandemic could definitely crash the world as we knew it. Nonetheless, we have never thought of the pandemic as some kind of eraser, a sort of ground zero from which to start a new world. Neither did we see it as some sort of revenge of nature on humans. Without venturing into scientific explanations of the causes of the pandemic, we argue that COVID-19 was the result of a long global history of unjust socio-ecological relations, and that it reinforced them rather than dismantling them. The lockdown was obviously an obstacle to social mobilization, although in some cities people organized solidarity brigades to support their most vulnerable community members (Armiero, 2021). The pandemic affected our academic project, Occupy Climate Change!, by slowing it down dramatically, but the main intuitions of that project could not have been more timely and correct. Between a global pandemic and a war, we were looking for the seeds of alternatives to the present state of affairs. In the face of gigantic, global problems, we explored the (counter-)power of community-led experiments, and in doing so we challenged the triad of mainstream climate change narratives: the comfortable future apocalypse, the scale of intervention and the post-political consensus. Climate change is always thought of in the future tense, something that will arrive as a biblical apocalypse, but, as with all prophecies, that does not scare us now. Things look different when seen through the eyes of communities on the front lines of climate breakdown, whether they are the indigenous populations of the Pacific Islands, working-class communities who cannot afford air conditioning, or migrants escaping from poverty or wars, but also droughts or floods. Solutions to climate change are either expected from nation-states and sets of states or they come down to individual choices. In other words, the politics of climate change seem to be made either at the COPs or at the supermarket. Instead, we have been working to recover both the power of collective action and community experiences that are happening everywhere. Erik Swyngedouw (2011) argues that climate change has become the ultimate post-political discourse/device: a space of hyper-consensus where the only thing needed is to follow the guidance of the experts within an unchallenged neoliberal socioeconomic paradigm. Although we do agree that acknowledging the science concerning climate change is a fundamental step, it still leaves plenty of room for diverse, antagonistic political options. Without a strong political stance concerning social justice and democracy, climate change measures can worsen conditions for subaltern communities and nonhumans (one can think of REDD+ schemes, gated communities and gentrification, and taxes that hit commuters rather than fossil fuel corporations). Looking at grassroots and local innovations, we explore the variety of diverse options that are emerging on this stage, and at times, they are deeply antagonistic to top-down approaches to climate change.

Repoliticizing climate change while shifting its temporal and scalar dimensions was at the core of our 2020 workshop and even more so of our three-year interdisciplinary collaborative research project, Occupy Climate Change! (hereafter, OCC!). This project scrutinized urban grassroots responses to climate change adaptation and loss and damage in five cities: New York City (USA), Rio de Janeiro (Brazil), Istanbul (Turkey), Naples (Italy) and Malmö (Sweden). In March 2020, the OCC! team hosted an international workshop in Stockholm in order to expand the scale of the analysis thematically and geographically. Building upon the OCC! project, the workshop focused on two key themes that have not been studied sufficiently in the literature on climate change: the social dimensions of loss and damage in cities and the roles played by grassroots community organizations in contesting loss and damage and in radicalizing climate politics.

The general framework of this volume draws from the OCC! project and its 2020 workshop in Stockholm. It builds upon critical approaches in political ecology and environmental humanities, advancing theoretical understandings and empirical studies of civil society engagement with climate breakdown, climate adaptation, and loss and damage. Recent works on climate change adaptation increasingly refer to the limits of adaptation, which denotes a cutoff line where loss and damage become inevitable. Translating the global debate on loss and damage as if people matter (Tschakert et al., 2017) requires attention not only to national, but also to local scales (Roberts & Pelling, 2018). Consequently, the exploration of adaptation as a process and not solely as an outcome, and of resistance as a manifestation of urban climate justice is at the heart of the contributions of this book. Studies of new climate urbanism (Long & Rice, 2019; Robin & Castán Broto, 2021), urban climate justice (Rice et al., 2023), urban political ecology (Kaika et al., 2023), do-it-yourself (DIY) adaptation (Cloutier et al., 2018), the Transition movement (Feola & Nunes, 2014), direct action (Vandepitte et al., 2019; De Rosa, 2022) and multiscalar and multisited politics of climate change all attest that, despite increasing interest in the actions of local governments and nonstate actors, grassroots mobilizations still receive scant attention (Chu, 2018). Moreover, contributions from the environmental humanities have also enriched this policy-oriented scholarship with a wide range of new perspectives and different layers (Gladwin, 2017; Dawson, 2017; Solnit, 2010). This book brings together contributions from both early career and established scholars to the broadly defined field of urban climate justice, scholars who dare to look beyond the myth of self-regulating markets of private insurance schemes and the liberal technocratic functionalism of engineering interventions. As such, it seeks to amplify grassroots voices from the Global South and Global North alike on issues including, but not limited to, radical adaptation (Dawson, 2017), bottom-up citizen initiatives (Shi et al., 2016), heterotopias (Edwards & Bulkeley, 2018) and climate gentrification (Anguelovski et al., 2016).

Positioning grassroots innovation at the center of climate action

When we began to develop our research project on climate change in 2016, Greta Thunberg was not yet sitting outside the Swedish parliament, and climate change seemed to be a specter in the political discourses of most European countries. Apart from some notable exceptions, the European press was also quite oblivious of climate change. Extinction Rebellion would be established only in 2018, and Friday was just the last day of the school week, with no special connotation. Nonetheless, it was clear to us that climate change was opening new spaces of mobilization both at the institutional and grassroots scales. The observation of empirical cases of extreme events demonstrated the ability of grassroots organizations and/or municipalities to counteract and experiment with self-organization and progressive solidarity initiatives (Harrington & Cole, 2022). Occupy Sandy, studied by Ashley Dawson, is a handbook case of this kind of self-organized community rescue. As Dawson (2017, p. 236) argued in his book, *Extreme Cities*:

> Disasters can often further strengthen capitalism and profit the rich, but they can also offer a glimpse of what radical political theorist Jodi Dean calls the communist horizon, the sense that the oppressive conditions of the present can be overcome and new forms of solidarity discovered.... Communal solidarities forged in the teeth of calamity can be seen as a form of disaster communism, under which people begin to organize themselves to meet one another's basic needs and to collectively survive.

Rebecca Solnit dedicated her volume *A Paradise Built in Hell* (2010) to the many stories of grassroots self-organization and solidarity born from the ruins of disasters, such as the San Francisco earthquake and fire, Hurricane Katrina in New Orleans, and the 1985 earthquake in Mexico City. The mainstream narratives that recount post-apocalyptic scenarios usually describe the dissolution of any kind of social bonds. Invariably, the science fiction industry proposes gangs of dehumanized survivors fighting with each other over the very few resources left. Such anti-social scenarios are projected beyond science fiction into the real aftermath of disasters, when the media amplify cases of loitering, and legitimize the imposition of harsh repression and states of siege. The historical cases of San Francisco and New Orleans instantiate this argument. We could add here that, even during the shipwreck of the *Titanic*, it was reported that some officers started shooting—allegedly in the air—to restore order among the third-class passengers who were trying to reach the lifeboats. Just to give a sense of the kind of narrative infrastructure supporting such an action, here is the statement of the officer who used his gun:

> I saw a lot of Italians, Latin people, all along the ship's rails—understand, it was open—and they were all glaring, more or less like wild beasts, ready to spring. That is why I yelled out to look out, and let go, bang, right along the ship's side.[1]

1 https://www.titanicinquiry.org/USInq/AmInq05Lowe07.php.

The *Titanic* is the perfect metaphor of the unjust impact of disasters—since on the *Titanic* class mattered very much indeed (Armiero, 2018). This brief statement by the law enforcement officer aboard the *Titanic* also reveals the clash between the imposition of an ordered reorganization of existing injustices and the attempt to sabotage them. Restoring normality is not what a progressive agenda looks like. For those studying disaster communism (Out of the Woods Collective, 2020), the rupture of the status quo opens to possibilities for practicing and imagining alternative ways of being a community. The very fact of exchanging outside the market and acting on motivations beyond individual profit is a radical challenge to the capitalist imperative.

Climate change, though, is a quite special disaster. It is ubiquitous not only in space, but also in time. It is occurring right now, but it has a strong projection into the future—the apocalypse is always a prophecy—and a shadow from the past. As several scholars have stressed, for indigenous people, the apocalypse is a memory, not a prophecy (Horne, 2018; Whyte, 2017). It becomes present and visible through dramatic epiphanies like Hurricane Katrina or the recurrent Australian wildfires, but most of the time it is invisible, especially to the rich living in the Global North. Nonetheless, we believe that some forms of disaster communism may also flourish when the disaster is not overt, but underground, ready to blow at any moment. Clearly, not all grassroots adaptation or mitigation strategies should be considered forms of disaster communism. In the introduction to a special journal issue about our work for the OCC! project, we proposed the category of "urban climate insurgency," which we describe as:

> the ensemble of grassroots initiatives that aim to tackle climate change from a radical point of view and take the city as the primary locus of action in doing so. In our vocabulary, insurgent does not imply violence but rather refers to the radical rejection of the current socio-ecological system. Urban climate insurgency does not follow the rules of the game; it does not legitimize the current climate regime through the paraphernalia of participatory tools that are designed to anesthetize anger and social mobilization.... It is insurgent because it clashes with mainstream climate policies, acknowledging that the climate crisis is not a mistake of the system but is the evidence that the system is deeply rotten and must be changed. (Dawson et al., 2022, p. 1)

In concrete, down-to-earth cases, the thousands of grassroots experiments tackling climate change often blend insurgent with less radical practices.

One can think of urban gardening projects, sometimes started through the illegal occupation of abandoned lands and then institutionalized by municipal governments. Even municipal governments assume, although in rare cases, an almost insurgent posture when, for instance, they disobey national directives by fostering an alternative agenda concerning climate change or migration. In this sense, we have decided to keep our net wide open. Activists on the ground may decide to collaborate with local institutions or to confront them. In both cases, we aim to explore the potentially radical contribution of their innovations.

Smith and Seyfang (2013, p. 384) attempted to translate the concept of innovation from the sphere of technology and business to that of civil society by showing how grassroots innovations can lead to "solutions that respond to the local situation and the interests and values of the communities involved." Not unlike business innovations, social innovations also involve the problem of diffusion, which can happen, according to Smith and Seyfang (ibid.), through the replication of small initiatives, the growth in scale of the initial project or the translation of pioneering ideas into mainstream settings. However, while in business innovations, these three steps are pushed by a quest for higher profits and market supremacy, in social innovations, the incentives must be found in ethical and/or political values, though sometimes seasoned with competition between different institutional actors—one could point to the international rankings of the most livable or green cities as a case in point.

Two features are crucial in the making of both grassroots and business innovations: creativity and knowledge. We argue that creativity in grassroots innovations is closely related to utopian thinking. Following Friedmann (2002, p. 3), we argue that utopian thinking plays a crucial role in imagining different futures beyond the limitations of what is possible in the present. According to Pieterse (2008), the production of knowledge, and thereby of new imaginaries, occurs through what he calls "epistemic communities." These communities assemble and network collective knowledge with the aim to "challenge conventional orthodoxy (the mainstream) about what is possible and not possible in terms of transformative urban development agendas" (ibid., p. 149). We believe that without a utopian agenda, even reformist policies would not be possible. Understanding the dialectic between utopian thinking and practical experiments is fundamental for any attempt to make sense of grassroots innovations in the field of climate change.

The primacy of knowledge production in the making of grassroots innovations is not surprising. After all, in the literature on business innovation, know-how is key. Can we actually distinguish between knowledge at the service of business innovation and the knowledge mobilized for grassroots innovations? As an initial attempt to delineate the main divergences between the two, we came up with five pillars of grassroots innovation (openness, multidisciplinarity, gratuity, place-basedness and collectiveness) as Table 1.1 shows.

Table 1.1. Knowledge for business innovation (K4BI) versus knowledge for grassroots innovation (K4GI)

K4BI	K4GI
Secrecy (in order to be profitable, knowledge leading to innovation must be kept private)	Openness (in order to foster social innovation, knowledge must be widely available)
Specialization (in general, although not always, business innovations require highly specialized knowledge)	Multidisciplinary (grassroots innovations are often produced through nonspecialized knowledge)
Marketability (at least potentially, knowledge for business innovation should have a market)	Gratuity (knowledge for grassroots innovation is not interested in the possibility of being sold)
Universalism (at least in principle, business innovations should be born from universal knowledge)	Place-based (in most cases, grassroots innovations are generated within community-rooted knowledge)
Traceability (it should be possible to attribute knowledge for business innovation to an individual researcher or at least to a team)	Collectiveness (it is often quite difficult to attribute a grassroots innovation to a specific researcher or team)

We argue, then, that these five key pillars (openness, multidisciplinarity, gratuity, place-basedness and collectiveness) are the foundational characteristics of grassroots innovation. This does not imply that in order to acknowledge and mobilize grassroots innovations, all those boxes must be ticked. Nor do we mean to imply that ticking these boxes leads necessarily to a progressive or even revolutionary agenda. Actually, mainstream discourses on climate change have begun to stress the relevance of local-level-centered policies and community resilience. In the general dismantling of state responsibilities, shifting duties to local authorities, without balancing this with more power and resources, seems the perfect move to shore up austerity policies rather than worry about the planet (See also Goh, 2020; Steele et al., 2015).

The image of the resilient community/subject is another recurrent trope in the neoliberal version of bottom-up climate policies. Radical scholars have voiced their critique of the resilience narrative. In a provocative article with the title, "'Don't Call Me Resilient Again!'" (2017), Maria Kaika argues that, instead of celebrating the resilience of subaltern communities, it would be much more fruitful to investigate the causes of the crises that have forced them to be resilient. Although we do acknowledge that resilience can be evoked in diverse and more progressive ways, we still argue that in its essence the capability to return to the state which was in place before the crisis diverges from a radical project aiming to address the causes of the crisis. In brief, it is neither the scale (city/community) nor the direction (bottom up) that makes a policy revolutionary or at least radically transformative. Rather, it is its ambition to reverse injustices, build equality and experiment with noncapitalist practices. Grassroots resistance movements (Temper et al., 2018) and social infrastructures of adaptation (De Rosa et al., 2022) have a clear role to play in such transformations.

We use the expression "Occupy Climate Change" to refer not so much to the historical "Occupy" movement, which developed mainly in 2011, but to the wider practices of reclaiming and managing spaces under noncapitalist logics. We are aware of and sympathetic with some indigenous critiques of the Occupy movement and of the very use of the word "Occupy" and its colonial legacy. Nonetheless, we claim the possibility to think outside the anglophone empire and its signification of everything. In many places around the world, "Occupy" refers to a completely different set of experiences and meanings. We are referring here to a diverse array of experiences, from workers' self-managed factories in Argentina to the agro-ecological communities of the Sem Terra movement in Brazil, from the ZAD areas in France to the squatter social centers of Italy. As in those cases, "occupying climate change" is, then, a liberation struggle which, while fighting against injustice, produces prefigurative politics. As Brownhill et al. (2022, p. 2) stress in their introduction to *The Routledge Handbook of Ecosocialism*, "capitalism's successor society is emerging right now, among us, in a multitude of prefigurative efforts, involving myriad moments and movements of transformation, building, and reinforcing the foundations of ecosocialist alternatives globally." Climate change discourses must be liberated from depoliticizing arguments, techno-fix solutions and false conscious universalism in order to liberate revolutionary potentials, which in the occupied spaces are prefigurative politics, suspended between utopia and experimentation. As such, this volume is our humble contribution to "the realisation of the apocalyptic reality of the ruinous now" that Swyngedouw (2023, p. 23) identifies clearly from which "a new politics might emerge. It is from within the ecological wreckage of the present that a new imaginary of the possible might arise."

From Naples to Honolulu: OCC! everywhere

Vanesa Castán Broto opens this volume with a meditation on hope in the context of climate change. While presenting the possibilities of disruptive innovations, she argues for adopting a reparative approach, that is, in her words, "the act of 'making good,' whether it is putting a broken object back to use or salvaging a broken relationship." The focus on reparative forms of climate action allows the author to acknowledge the "historical legacy of exclusion and oppression" that, when not considered at all, can negatively affect even the most positive social innovations. All the following chapters, moving from radically diverse geographies and disciplinary backgrounds, deal with the intersection between climate vulnerability and social injustices.

Grassroots struggles are at the core of all the case studies presented in this volume, although their articulation is often rather diverse as, for instance, in the cases of New York City's mobilization to ensure public ownership of electric power (Dawson) and Hawai'ian opposition to state-led adaptation measures (Huet). Clearly, the authors reject the mainstream tactic of separating the "good" activists (those proposing solutions) from the "bad" activists (those resisting injustice while trying to stop harmful policies). In New York, for instance, activists mobilized to stop fracking while proposing a model for publicly owned energy providers. Cristina Visconti shows that in Chile the almost insurrectional strategy of the protestors was coupled with experiments of social innovations such as *ecobarrios*. This implies that social innovations—and the knowledge and practices that build them—are generally produced through socio-environmental struggles. In this sense, we believe that socio-ecological conflicts can be part of the solutions if we stop looking for top-down, imposed techno-fixes.

Several authors in this volume agree with bell hooks' intuition of the revolutionary—or at least transformative—power of the margins. Berruti and Palestino, Sedrez and Barbosa, and Henrique all present cases of marginal communities—in Naples (Italy), Rio de Janeiro and São Paulo (Brazil), respectively—that have developed creative forms of environmental activism producing new subjectivities while striving to improve the material conditions of life. In these three cases, the reclamation of blighted places merges with a socio-ecological project aiming to create a nurturing community built on solidarity and sharing. The articulation between grassroots innovations and institutional support is a significant challenge in these chapters as well as across this entire volume. Sometimes the fragility of the institutions does not allow for any meaningful interaction (Berruti and Palestino). In any case, the capability to affect institutional politics and policies does not depend on the degree of radicalism deployed by social movements—assuming that more radicalism implies less concrete results in the interaction with institutions. Dawson, for instance, points at the connections between the grassroots mobilizations on climate justice and the elaboration of the Green New Deal. Moving from a characterization of the consensual regime of climate change governance as a manifestation of hegemony, De Rosa finds possibilities for transformative openings in grassroots and activist strategies of repoliticization that bring from the margins to the center knowledges and experiences that unsettle technocratic and market-based solutions. In some cases, these have led to successful outcomes, as with the blockade of the gas terminal in Göteborg and the Swedish government's decision to halt its expansion.

In all the cases presented in this volume, the development of specific forms of climate actions implies the production of transformative knowledge. In some cases, this takes on the appearance of local knowledges that are rediscovered and reinterpreted, for instance, the farming culture of the Ponticelli neighborhood in Naples. Other times, it implies the performative invention of communities, as in the activists' camps and urban social centers analyzed by De Rosa. For instance, Kotsila reflects on the untapped and actually dismissed knowledge produced within migrants' communities in Europe. As she writes, "critical scholarship should treat immigrants as equal knowledge-holders, with experience, expertise and situated—embodied, partial and accountable—knowledge(s), who can offer crucial insight when it comes to reconceptualizing and planning for healthy and just cities." Boda, on the other hand, discusses the challenges of connecting scientific knowledge and environmental mobilization through a case study of beach protection in a small town in Florida. As Boda writes,

> One fruitful way forward may be to explore more radical forms of collaboration between civil society organizations and scientific knowledge producers, such that the tactical and other forms of experiential knowledge, so essential to movements, can be combined with objective knowledge of economic and political structures.

Artist and activist-scholar Aurash Khawarzad closes the volume with a clear invitation to action. His chapter should be read as an instruction booklet on how to produce a people's climate plan in every community. The chapter derives from the experience that its author and Ashley Dawson have acquired by producing *A People's Climate Plan for New York City?* with the OCC! project (Climate Action Lab, 2019). Although the chapter offers some rather specific suggestions, Khawarzad rightly insists that the people's climate plan is not a set of rules or steps to follow, but a process that leads to creating an empowered community.

Perhaps this is true for the entire OCC! project, and thus for this volume as well. The diverse and multifaceted experiences we have collected through the project and the few offered in this volume are not exemplary in the sense of providing sets of rules, or the template, for "the perfect radical climate initiative." Every case is place-based and context specific. The point is not to set up a canon—practically an oxymoron when speaking of radical activism—but to exercise the revolutionary practice of listening to stories that generally remain invisible. Thousands of small acts of resistance happening now are waiting to be heard, and if the process is as important as—or perhaps even more than—the results, learning to listen and notice is not only a research strategy, but an integral part of the transformation process.

References

Anguelovski, I., et al. (2016). From toxic sites to parks as (green) LULUs? New challenges of inequity, privilege, gentrification, and exclusion for urban environmental justice. *Journal of Planning Literature*, *31*(1), 23–36. https://doi.org/10.1177/0885412215610491

Armiero, M. (2021). *Wasteocene. Stories from the global dumps.* Cambridge University Press.

Armiero, M. (2018). Sabotaging the Anthropocene, or, Praise for mutiny. In G. Mitman, M. Armiero, & R. Emmett (Eds.), *Future remains: A cabinet of curiosities for the Anthropocene* (pp. 129–139). University of Chicago Press.

Brownhill, L., Engel-Di Mauro, S., Giacomini, T., Isla, A., Löwy, M., & Turner, T. E. (2022). Preface: An introduction to ecosocialism. In L. Brownhill, S. Engel-Di Mauro, T. Giacomini, A. Isla, M. Löwy, & T. E. Turner (Eds.), *The Routledge handbook on ecosocialism* (pp. 1–12). Routledge.

Chu, E., Anguelovski, I., & Roberts, D. (2017). Climate adaptation as strategic urbanism: Assessing opportunities and uncertainties for equity and inclusive development in cities. *Cities, 60*, 378–387. https://doi.org/10.1016/j.cities.2016.10.016

Climate Action Lab. (2019). *A people's climate plan for New York City?* The Graduate Center, City University of New York. https://occupyclimatechange.net/wp-content/uploads/2022/03/ClimatePlan_FINAL1-1.pdf

Cloutier, G., Papin, M., & Bizier, C. (2018). Do-it-yourself (DIY) adaptation: Civic initiatives as drivers to address climate change at the urban scale. *Cities, 74*, 284–291. https://doi.org/10.1016/j.cities.2017.12.018

Dawson, A. (2017). *Extreme cities: The peril and promise of urban life in the age of climate change*. Verso.

Dawson, A., Armiero, M., Turhan, E., & Biasillo, R. (2022). Urban climate insurgency: An introduction. *Social Text, 40*(1), 1–20.

De Rosa, S. P. (2022). Breaking consensus, transforming metabolisms: Notes on direct action against fossil fuels through urban political ecology. *Social Text, 40*(1), 135–155.

De Rosa, S. P., De Moor, J., & Dabaieh, M. (2022). Vulnerability and activism in urban climate politics: An actor-centered approach to transformational adaptation in Malmö (Sweden). *Cities, 130*, 103848.

Edwards, G. A. S., & Bulkeley, H. (2018). Heterotopia and the urban politics of climate change experimentation. *Environment and Planning D: Society and Space, 36*(2), 350–369. https://doi.org/10.1177/0263775817747885

Eriksen, S. H., Nightingale, A. J., & Eakin, H. (2015). Reframing adaptation: The political nature of climate change adaptation. *Global Environmental Change, 35*, 523–533. https://doi.org/10.1016/j.gloenvcha.2015.09.014

Diouf Sarr, M. (2022, November). At COP27, support poorest for climate loss and damage. *Nature, 611*, 9. https://doi.org/10.1038/d41586-022-03474-1

Feola, G., & Nunes, R. (2014). Success and failure of grassroots innovations for addressing climate change: The case of the Transition movement. *Global Environmental Change, 24*, 232–250. https://doi.org/10.1016/j.gloenvcha.2013.11.011

Friedmann, J. (2002). *The prospect of cities*. University of Minnesota Press.

Gladwin, D. (2017). *Ecological exile: Spatial injustice and environmental humanities*. Routledge.

Goh, K. (2020). Urbanising climate justice: Constructing scales and politicising difference. *Cambridge Journal of Regions, Economy and Society, 13*(3), 559–574.

Harrington, E., & Cole, A. (2022). Typologies of mutual aid in climate resilience: Variation in reciprocity, solidarity, self-determination, and resistance. *Environmental Justice, 15*(3), 160–169.

Horne, G. (2018). *The apocalypse of settler colonialism: The roots of slavery, white supremacy, and capitalism in seventeenth-century North America and the Caribbean*. Monthly Review Press.

Kaika, M. (2017). "Don't call me resilient again!": The New Urban Agenda as immunology … or … what happens when communities refuse to be vaccinated with "smart cities" and indicators. *Environment and Urbanization, 29*(1), 89–102. https://doi.org/10.1177/0956247816684763

Kaika, M., Keil, R., Mandler, T., & Tzaninis, Y. (Eds.). (2023). *Turning up the heat: Urban political ecology for a climate emergency*. Manchester University Press.

Le Quéré, C., Jackson, R. B., Jones, M. W., et al. (2020). Temporary reduction in daily global CO_2 emissions during the COVID-19 forced confinement. *Nature Climate Change, 10*, 647–653. https://doi.org/10.1038/s41558-020-0797-x

Long, J., & Rice, J. L. (2019). From sustainable urbanism to climate urbanism. *Urban Studies, 56*(5), 992–1008. https://doi.org/10.1177/0042098018770846

Masterson, V. (2022). How has the war impacted Ukraine's environment? World Economic Forum. https://www.weforum.org/agenda/2022/07/ukraine-war-environmental-impact/

Out of the Woods Collective. (2020). *Hope against hope: Writings on ecological crisis*. Common Notions.

Pieterse, E. (2008). *City futures: Confronting the crisis of urban development*. Zed Books.

Rice, J. L., Long, J., & Levenda, A. (2023). *Urban climate justice: Theory, praxis, resistance*. University of Georgia Press.

Roberts, E., & Pelling, M. (2018). Climate change-related loss and damage: Translating the Global Policy Agenda for national policy processes. *Climate and Development, 10*(1), 4–17. https://doi.org/10.1080/17565529.2016.1184608

Robin, E., & Castán Broto, V. (2021). Towards a postcolonial perspective on climate urbanism. *International Journal of Urban and Regional Research, 45*(5), 869–878. https://doi.org/10.1111/1468-2427.12981

Shi, L., Chu, E., Anguelovski, et al. (2016). Roadmap towards justice in urban climate adaptation research. *Nature Climate Change, 6*(2), 131–137. https://doi.org/10.1038/nclimate2841

Smith, A., & Seyfang, G. (2013). Constructing grassroots innovations for sustainability. *Global Environmental Change, 23*(5), 827–829. https://doi.org/10.1016/j.gloenvcha.2013.07.003.

Solnit, R. (2010). *A paradise built in hell*. Penguin.

Steele, W., Mata, L., & Fünfgeld, H. (2015). Urban climate justice: Creating sustainable pathways for humans and other species. *Current Opinion in Environmental Sustainability, 14*, 121–126.

Swyngedouw, E. (2011). Depoliticized environments: The end of nature, climate change and the post-political condition. *Royal Institute of Philosophy Supplement, 69*, 253–274. https://doi.org/10.1017/S1358246111000300

Swyngedouw, E. (2023). "The apocalypse is disappointing": Traversing the ecological fantasy. In H. Haarstad, J. Grandin, K. Kjærås, & E. Johnson (Eds.), *Haste: The slow politics of climate urgency* (pp. 17–25). UCL Press.

Temper, L., Walter, M., Rodriguez, I., Kothari, A., & Turhan, E. (2018). A perspective on radical transformations to sustainability: Resistances, movements and alternatives. *Sustainability Science, 13*(3), 747–764. https://doi.org/10.1007/s11625-018-0543-8

Tschakert, P., et al. (2017). Climate change and loss, as if people mattered: Values, places, and experiences. *WIREs Climate Change, 8*(5), e476. https://doi.org/10.1002/wcc.476

Vandepitte, E., Vandermoere, F., & Hustinx, L. (2019). Civil anarchizing for the common good: Culturally patterned politics of legitimacy in the climate justice movement. *VOLUNTAS: International Journal of Voluntary and Nonprofit Organizations, 30*(2), 327–341. https://doi.org/10.1007/s11266-018-00073-5

Whyte, K. (2017). Indigenous climate change studies: Indigenizing futures, decolonizing the Anthropocene. *English Language Notes, 55*(1–2), 153–162.

WMO. (2022). *United in science 2022: A multi-organization high-level compilation of the most recent science related to climate change, impacts and response.* World Meteorological Organization. https://library.wmo.int/doc_num.php?explnum_id=11308

About the authors

Marco Armiero is ICREA Research Professor, Institut d'Història de la Ciència (IHC), Universitat Autonoma de Barcelona, Spain. From 2019 to 2023 he served as the president of the European Society of Environmental History. Formerly, he directed the Environmental Humanities Laboratory (EHL) at the KTH Royal Institute of Technology in Stockholm, Sweden. He has extensively published on environmental justice, climate change, migration and the nationalization of nature. He is the editor-in-chief of Resistance. A Journal of Radical Environmental Humanities.

Salvatore Paolo De Rosa is a researcher at the Center for Applied Ecological Thinking, University of Copenhagen. Previously, he was a researcher at Lund University in Sweden, where he also received his PhD in human geography. With a background in anthropology and political ecology, his research focuses on collective action and socio-environmental transformations towards sustainability and justice.

Ethemcan Turhan is an assistant professor of environmental planning at the Department of Planning and Environment, University of Groningen. His research is situated in the broadly defined field of political ecology with empirical attention to environmental conflicts, energy infrastructures, social movements, and climate mobilities mainly (but not exclusively) in Turkey, the larger Middle East, and the Mediterranean. He is the coeditor of the book *Transforming Socio-natures in Turkey: Landscapes, State and Environmental Movements* (Routledge, 2019).

2. Hope in something: An earthly tragedy in five acts

Vanesa Castán Broto

Abstract

This chapter explores the generative potential of the unexpected and the experimental in bringing about transformational change. By reflecting on works of art, personal experiences and scientific literature on climate action and adaptation, Castán Broto brings to the fore the tensions between hope and the compulsion to act, between sense of responsibility and paralysis in the Anthropocene, between disruption and innovation. The demands for rapid changes at scale, while justified in their urgency, risk to lead to dilemmas concerning energy transitions and the issue of who will have to pay for such disruptions. Therefore, an alternative form of reading the crises should move away from a paranoid one toward building a reparative ethos. Castán Broto concludes by proposing reparative climate action.

Keywords: climate change adaptation, reparation, hope, loss, responsibility

ACT I: In which we revisit ideas of immobilism in the Anthropocene

Some books irrupt into your life and force their material presence into your thinking. This is what the beautiful book *Critical Zones: The Science and Politics of Landing on Earth* (Latour & Weibel, 2020) did for me: first, I became obsessed with it, then it shifted my viewpoint. The book arrived by mail, and I was surprised by its size and weight. The cover immediately captured my imagination: I loved the texture of the cover's cardboard, and I was intrigued by Sophie Ristelhueber's photograph, *Sunset Years #2*. It is a picture of craters and disruptions on a mud surface, drying in the sun.

Armiero, M., E. Turhan & S. Paolo De Rosa (eds), *Urban Movements and Climate Change: Loss, Damage and Radical Adaptation*. Taylor & Francis Group 2023
DOI: 10.5117/9789463726665_CH02

It made me think of a mud metaphor I wrote with my colleagues, Harriet Bulkeley and Gareth Edwards, about the role of experiments in social change:

> Within socio-ecological and socio-technical networks, an experiment acts as a source of disturbance which is not confined to a particular place and moment, but rather reverberates through the network with the potential to reconfigure its constituent elements and the circulations of which they are part. An illustration might be found in that of a boiling viscous liquid, such as the mud pools found in Rotorua, New Zealand, where bubbles of air not only disturb the surface but create new patterns, imprints of their passing, which in turn shift the surface topology, creating new patterns, creating new forms of potential. (Bulkeley et al., 2014)

In this quote, we highlight the generative potential of the unexpected. Even tiny bubbles can change the trajectory of such a viscous liquid. That is how social change looked to me at the time: a viscous flow that is impossible to apprehend or control in its totality, where disturbances within the flow determine movement trajectories. This metaphor captured the dynamics of change we observed in cities, where diverse actors—both local government and citizen-led movements—were actively assembling intervention spaces. Or where climate action could form those tiny bubbles that could eventually bring about a transformational change in a society stuck in its old, polluting ways. In a contribution to the edited volume *Innovating Climate Governance*, Harriet and I picked up on the bubbling mud metaphor to analyze the transformative power of experiments through their circulation in urban environments:

> [T]he transformative potential of the experiment is always and already part of the intervention, and its capacity for transformation relates both to the "stickiness" of the regime within which it is inserted and its immanent capacity to assemble, enroll and transform the socio-technical configurations of which it is a part. (Castán Broto & Bulkeley, 2018, p. 75)

These reflections informed the conclusion to the volume, whose editors examine the multiple forms of embedding experiments, including the idea of "circulation" (Turnheim et al., 2018). Circulation is tied here both to a poetic vision of experiments flowing through a "governance milieu" and to a reimagination of experiments as "journeys" (see Carvalho & Lazerini, 2018; Pallet, 2018; Heiskanen & Matschoss, 2018). Action is about movement, about shifting things around. Yes, we do not always know what to do. Should that paralyze us? No: we can agitate, bring opportunities together, demonstrate what may be possible. We can experiment, both with our lives and with the systems of governance that structure them.

Before I read *Critical Zones*, I had critical questions about experiments, but this large, imposing book forced me to look those questions in the eye. Cracked patterns develop around the craters as the mud dries. These photographs of drying mud,[2] of craters and cracks congealed in it, reflect a state of immobility, becoming even more immobile because of the drying effect of the sun. Looking at this photo, I felt the liquid mud metaphor no longer spoke of shared directions. Does it matter? It does because hope has emerged around the idea of working via the intervention spaces afforded by the mundane and their promise of international impact (Acuto, 2014). These drying mud pictures force a reckoning with the problem of paralysis, the realization that what has been done is far too little and too late.

Latour mocks the current search for an "ecological cure" for the Anthropocene as something that ignores that "there is no cure for the condition of belonging to the world" (Latour, 2017, p. 13). Hope is dangerous, Latour argues. It is better to hope for the time when counting on hope is no longer the most reasonable action. It is indeed a worrying fact that many climate scientists repeat the mantra, "Everything is doomed, but I still have hope." This is a mantra that invokes the possibility of a surprise. Such surprises are never too far from miraculous solutions (geoengineering anyone?), and miraculous solutions are never too far from a swindle. Perhaps this time is, indeed, the time when our hope will serve us no longer—no further than it does the drying cracks in Ristelhueber's photos.

The Whitechapel Gallery in London has a biography of Ristelhueber on its website. It describes her work as a reflection on territory and its history in which "she strives to implement the bare act and the stamp of history on both the body and on the landscape, by rendering visible wounds and scars, veritable memories of the 'acts' of history" (Whitechapel Gallery, 2017). The whole series of photos depicts "climate bombs": craters in the Dead Sea illustrating the devastation caused by the overuse of water from the River Jordan and craters in the asphalt in Paris, macro and micro symptoms of a world in turmoil (L'Œil de la Photographie, 2019). Metaphors of the impasse in climate action. Congealed future memories of climate changed pasts. The photos put the idea of dynamism through bubbling experiments to sleep, drawing a giant question mark in their place. What can we do? Is this the anteroom to paralysis?

2 There is another similar photograph, *Sunset Years #5*, on the back of the book.

ACT II: In which a bird in a storm becomes a symbol for a certain compulsion to act

A book is not its cover, of course. *Critical Zones* offers a journey through multiple ways of confronting, accepting and embracing the Anthropocene. As I was thumbing through it, admiring its illustrations, the book fell open to the essay "The Point of View of the Mountain" by art historian Estelle Zhong Mengual (2020). Her essay discusses Albert Bierstadt's monumental *A Storm in the Rocky Mountains, Mt. Rosalie* as a historical painting that portrays the landscape that already is and the landscape that it will become (due to the storm) (Figure 2.1). Zhong Mengual writes:

> The storm is an internal agent that blends into the mountain as it creates its forms and partially determines the life that can live on it. It is a slow, blind, and fine sculptor that works in tiny strokes to give form to it. Only the mountain has lived long enough to listen objectively to the howl of the storm. (Zhong Mengual, 2020, p. 253)

Alongside the essay, seven illustrations depict the picture and its details. My daughter sensed my excitement and came to ask me what I was reading, and I gave her one of those superficial explanations one offers when wanting to be left alone with a precious book. But my daughter insisted. "Can we take a moment to appreciate the brave little bird flying through the storm?" she said, laughing. And I laughed, too, although it was only then that I appreciated this detail of the painting. The bird is there, floating in the clouds. Once you see it, it is impossible to unsee it; everything in the painting refers to the brave little bird in the storm. There is something compelling about the understated presence of the bird in this painting. It generates many questions that move beyond the point of view of the mountain presented by Zhong Mengual: why is the bird flying as the storm approaches?

Is it escaping from the destructive grasp of the storm? My rationalist mind immediately assumes that the bird's flight has a clear, adaptive purpose. Ethologists have shown that some birds can predict changes in weather conditions using infrasonic cues or because of their sensitivity to barometric pressure. Birds may thus fly ahead of storms or even into them. In this reading, the bird becomes a metaphor for anticipatory action, working in syntony with nature, and having the intuition to respond effectively to the massive changes around us. However, we should remind ourselves that shelter in place is possibly a more common strategy among birds to weather storms.

Figure 2.1. *A Storm in the Rocky Mountains, Mt. Rosalie*, by Albert Bierstadt, 1866.

Of course, this is just a painting, not an actual storm, an actual mountain and an actual bird. Bierstadt may have only put the bird in the painting because he just saw it passing by. An accident immortalized. Nevertheless, it is a particularly odd occurrence that is not frequently repeated in his other paintings. Regardless of why the bird found its place in this painting, its flight is not one that can easily be read in consequentialist terms. My daughter used the words, "brave little bird," and I would suggest that her choice of words was not accidental. She did not see a bird that was clever, perceptive or efficient. Instead, the bird in the picture appears to engage in the futile action of traversing the storm. It represents a certain form of useless heroism that matches the romantic tones of the painting. Will something change because of the bird's flight? What will it change and for what purpose? Zhong Mengual's essay suggests that we must pay attention to the multiple biotic and abiotic processes woven into the history of a landscape for which the storm is no more than a heartbeat. However, the one thing that sets apart the bird in this landscape is its deliberate flight against the elements and its loneliness in the middle of the cloud. And the bird itself carries the weight of its history and all the places where it has flown.

The bird stands for me as a metaphor for the current moment of transition. Perhaps a new metaphor, in place of the lost one about the shifting mud. The bubbles in the mud are part of a broader process of change and create a sense of collective direction; the bird less so. The current need for change is most often imagined as a flight into the storm, anticipating the risks, being cleverer than the looming danger. I do not deny that this is sometimes how social change feels. However, significant changes are often made up of myriads of futile, daring actions whose actual impact may escape our senses. What is common to all those interpretations is a compulsion to act, a need to react and show movement—this is also an action that must be performed and made visible to count in the broader landscape of climate politics. In the face of existential fright, we may feel like little birds battling our wings in the storm. Is this a futile plight?

There is a certain contradiction embedded in an Anthropocene that recognizes humans as a geological force while also highlighting humanity's absolute loneliness and sense of impotence. It generates a sense of responsibility as much as it paralyzes us.

ACT III: In which possibilities of transition have turbulent consequences

Humanity is operating in crisis mode. The climate emergency has led to panicked responses. Declarations of emergency have swept institutions, following a youth-led social movement that has built momentum for radical transformation. Promises of radical transformation proliferate in academic discourse, both from the natural and social sciences. Transformation thinking seeps into policy and practice with promises of large-scale changes in how we live and how we understand ourselves—prescriptions for large infrastructure investment programs such as the Green New Deal, which in Europe has crystallized into a continent-wide plan for action, contain the promise of a government-led global transformation.

In February 2021, the Secretariat of the UN Framework Convention of Climate Change (2021)[3] published a synthesis of climate action ambition as contained in countries' new or updated nationally determined contributions (NDCs) up to December 2020. The message was clear: some progress, but not enough. For example, the report highlights that the quality of NDCs, including data on mitigation targets, has clearly increased. The implementation of the plans is addressed much more comprehensively, making linkages to relevant national planning, regulatory and legislative processes.

The synthesis report struggles to reconcile the estimated needs for carbon reductions with the promises embedded in the NDCs. Renewed calls for urgency have followed. These calls for urgency echo social movements,

3 This is how the UNFCCC's website refers to it. See: https://unfccc.int/news/updated-ndc-synthesis-report-worrying-trends-confirmed.

especially the youth movement Fridays for Future, and the spread of declarations of climate emergency to almost 2,000 governmental bodies worldwide. There is an apparent demand to shift our societies away from social and economic models that do not support a resilient and equitable future. The 2021 United Nations Climate Change Conference (COP26), held in Glasgow, Scotland, "requested" that parties improve their efforts in 2022, but there is skepticism about whether they will step up their efforts (Carbon Brief Staff, 2021).

These transformations, however, will necessarily be turbulent: evidence is mounting about how transformations exacerbate vulnerabilities and create misery by leading to arbitrary decisions about whose lives will be valuable enough to preserve (Schipper et al., 2021). Maladaptation is a ghost haunting climate adaptation (Eriksen et al., 2021). Disruptions and experiments drive transformations, but always at a price. Radical climate action can also be linked to a new politics of death or necropolitics (Mbembe, 2019), in which value is assigned to different segments of populations. Who you are determines whether you survive or not. Not surprisingly, many voices rage against the uncritical adoption of such assumptions in climate action.

Disruptions are integral to theories of transitions, a field that has influenced current transformations thinking. In studies of sustainability transitions, a transition happens when disruptive innovations find their way into the dominant regime, forcing a realignment of social and material components. These analyses are crucial because disruption can be seen as a "prerequisite of system reconfiguration," rendering the disruptive dynamic indispensable to transition processes (Kivimaa et al., 2021). Change is not always disruptive, but the rapid change required for a transformation of this kind will need disruption.

In business studies, disruptive innovations are those that alter a given market and technological context radically. Disruptive innovation was championed by economist Clayton Christensen, in contrast to sustaining innovations that introduce change incrementally. Disruptive innovations are low-cost, accessible technologies that produce sweeping and fundamental changes across markets. Disruptively innovative entrepreneurs and inventors become agents of change in industries and society. Their efforts establish new value propositions, change consumer behaviors and displace incumbent firms (King & Baatartogtokh, 2015, p. 77).

Building on this thinking, transition scholars argue that disruptive innovation fosters sustainability transitions. For example, renewable energy technologies are frequently framed as disruptive innovations that challenge fossil fuel-dependent production and consumption systems. Disruptive innovations in solar energy may include customer-oriented solutions for solar systems, but the range of disruptive innovations in the current infrastructure regime is varied, ranging from hydrogen fuel cells to sharing economy solutions. In summary, there is no question that disruptions drive large-scale changes. Transitions require disruption (Kivimaa et al., 2021).

The discourse of urgency and its translation into ideas of disruption raises eyebrows, especially among those concerned with the unequal distribution of vulnerabilities to climate impacts and climate action. In their recent edited volume on the dilemmas of energy transitions in the Global South, Ankit Kumar, Auke Pols and Johanna Hoffken write:

> While urgency is crucial for energy transitions in a climate-changed world, we need to be wary of haste. We must be cautious of conflating goals and processes of sustainable development and enquire what urgency means for due process. Justice needs thought, participation and deliberation.... Taking the space and time in which these transitions take place into account is critical in thinking through these dilemmas. (Kumar et al., 2021, p. 182)

Their words, emerging from concerns about the practical implementation of energy transitions in the context of deep, entrenched inequities, contrast with the demands from within social movements for rapid changes at scale. Transitions will take place at a price, and it is highly likely that this price will be paid by further marginalization and exclusion by those who are already most vulnerable. Not only do we need to be wary of the possibilities for social change embedded in experiments and disruptions, but we also need to foreground the potential negative impacts that such action can have on different vulnerable groups, incidentally groups who are rarely recognized as active agents in low carbon, resilient transitions, despite their visibility on the streets.

ACT IV: In which a reparative impulse provides the means to rethink low carbon transitions and climate action

Are there alternatives to disruptive innovation? In a recent paper, the LOACT[4] team (Castán Broto et al., 2021) argues in favor of thinking climate

4 LOACT (Low Carbon Action in Ordinary Cities) is the project that funded this research. See: https://www.loactproject.com/.

innovation differently through a "reparative" approach. Reparations are linked to the act of "making good," whether it is putting a broken object back to use or salvaging a broken relationship. Our primary source of inspiration is Eve Kosofsky Sedgwick (1997), who calls for a different reading of a broken world. As a cultural theorist, Sedgwick is mainly concerned with the act of reading texts and situations. Sedgwick's essay criticizes a particular form of critical reading that focuses on anticipating negativity, a paranoid reading of the world. For example, critical political economy perspectives on climate action often suffer from paranoia when they anticipate that any attempt at climate action will be doomed. Such readings take a privileged vantage point beyond the reader, a point of view that aims to anticipate surprises and foreground the dire state of circumstances to avoid being lost. Sedgwick refers to this point of view as critical omnipotence. What good is knowing better without a response? For this reason, such paranoid readings come close to attempts at solutionism and are tied to the politics of hope, whether hope resides in technology, political action or revolution.

Sedgwick suggests that an alternative form of reading world crises exists and argues that moving away from paranoid reading is the first step to building a reparative ethos. Here, reparation means engaging with how to make small worlds to provide sustenance. In this vein, a reparative reading is always tied to action. Discussing Sedgwick's work, Wiegman (2014) speaks of the need for reparative work because a reparative perspective is indissoluble from reparative action. Embracing the reparative is a way of responding with affirmative richness to every life encounter: it means allowing oneself to be surprised by the horror of life. Weighing in on this debate, Hanson (2011) proposes accepting a world that is damaged and dangerous and live within that world in an attempt to build alternative sets of relations.

The focus on "making amends" in a reparative reading relates to our relationship with others and other things. It requires recognizing that something is wrong, and that payback or compensation may be needed for those people and things that have been wronged. Theories of restoration and reparative justice can support a reparative reading of the climate crisis. Philosopher Margaret Walker (2010), for example, has argued that reparation implies remedial action to address the wrongs of the past. Walker explains that thinking about reparative justice is different from thinking about corrective justice. In the context of climate change, reparative justice calls for examining the history of climate change as a problem, as a means of assigning responsibilities for reparation. Climate change is a profoundly traumatic event, particularly considering human culture's continuity. Reparation will be a necessary but not sufficient part of healing from trauma.

By embracing reparative innovation in climate action, I seek to bring these insights into approaches that address human vulnerabilities through knowledge-making and innovation. To develop a reparatory reading of the climate crisis that, instead of rushing for disruptive action, seeks to engage with multiple means of making amends. When facing an uncertain world, the question is whether we can embrace a reparative impulse and develop positive attachments even within a heteronormative society that prioritizes the protection of some over others, that sees some lives as inherently more valuable. Against critical analysis that seeks to anticipate the hidden meaning of everything, reparative readings situate the object of knowledge within a material history of multiple experiences and attachments.

Confronting the trauma of the climate crisis also requires a reparative impulse that accepts the horrors of climate change and provides hopeful responses. One hopes that these reparative impulses are widespread, but in reality, most reparative responses have developed out of necessity. Transforming an impulse into an innovation depends on linking action and context in the hope of activating change. One of the best examples of reparative innovation I have encountered is the collaboration between researchers at Makerere University (Uganda) and urban communities in Kampala (Lwasa et al., 2020). Collective forms of waste management are essential when municipal waste collection may not reach the most disadvantaged settlements where waste accumulates (Lwasa, 2019). The innovation in question relates to briquette-making. Briquettes are compressed blocks of combustible materials. In this case, briquette-making is an opportunity to transform combustible waste into a household fuel, a commercial product that can support small businesses. In Kampala, communities organize themselves to sort their neighborhood waste to select the fraction that can be carbonized into briquettes (Kareem & Lwasa, 2011).

Briquette-making in Kampala is a reparative innovation because it involves, first of all, appropriating cityscapes by challenging accepted meanings, in this case, provoking instability in current management systems and rethinking what waste is. Briquette-making involves the collaboration of community-based organizations with university partners and other institutions that can frame this activity as innovation, so there is a tricky dependency between who is allowed and who is not allowed to produce innovative knowledge. Reparative innovation, however, requires collaborations that facilitate a process of urban reclamation from which communities can also benefit. Communities thus engage with the fundamental material characteristics of the waste. Such material agencies of objects are central to any process of social change (Liboiron, 2016). However, these material agencies cannot necessarily be controlled. For example, the communities in Kampala have raised concerns about pollution in the briquette-making process. That has motivated further innovation in their work with Makerere University, from the selection process of waste to the combustion methods used. Reparative innovation requires accepting uncertainty: uncertainty in political processes and knowledge attribution, but also material uncertainty.

Discarded resources can be mobilized for the community, attaching new instrumental value to discarded resources and spaces. Due to the lack of attention to such resources elsewhere, people facing exclusion gain access to resources. In doing so, they can build new markets. Briquette-making also transgresses boundaries between the public and the private by facilitating the development of a social economy that involves the collective production of a shared good that can be "privatized" to sustain the most disadvantaged groups within the community. Scholars at Makerere University have long worked with communities to understand responses such as briquette-making and foreground it in debates on sustainable urbanism, for example, in helping redefine the circular economy. I see their work as an example of how to engage practically with the realities of urban life in the lives of academic-activists: not by dictating the course of action, but by understanding the myriad ways in which people attempt to repair their lives. However, since the world's reserves of environmental injustices are inexhaustible, reparative innovation may call for scaling up, multiplying and accelerating action beyond existing capacities, in a call for infinite responsibilities that may eventually make people so tired that they give up (Dave, 2017). Naisargi N. Dave explains that the generation of infinite responsibility robs people of the will to persevere and calls for alternative immanent ethics that extracts determination and persistence not from the future, but from the present (ibid.).

In the end, reparative innovation stands in the boundary between social innovation (generating new solidarity networks) and technological innovation (mobilizing the material agency of waste). Reparative innovations may support social movements capable of claiming the community's political interests beyond a specific project by facilitating a mobilization process, but that is not why they take place. Like the bird flying in the storm, reparative innovation requires engaging with small, sometimes wild, possibilities. The assumption is that even the smallest actions will bring some form of urban change. In sum, reparative innovation endorses an alternative reading of the change process, one that recognizes radical change as the unexpected product of long-term, uncoordinated shifts.

In the infrastructural context of less-developed urban areas, reparative innovation calls for positive imaginations of collective urban futures that acknowledge difficult histories of colonial exploitation and resource extractivism.

ACT V: In which we explore alternative points of view to generate a new sense of hope

I called this chapter a tragedy because, even if reparative innovation can become a workable alternative to deliver just transitions, it generates more questions than answers: Who will lead reparative innovation? How will it happen, and with what consequences? The possibility of having unintended effects always hangs over innovation, whether we call it reparative or otherwise.

Many climate change experiments have a reparative component. Nevertheless, they face the indictment of being localized and not having enough impact. There is indeed a question about the extent to which reparative innovation will be an acceptable option in the context of urgency and what political possibilities may generate.

Reparative innovation is a means of examining what kinds of social innovation will lead to social change that will enable reparations to those who suffer the most—people and the world. As a provisional, workable alternative, one that engages with the practical possibilities for climate action, reparative innovation may work effectively as a counterpoint to disruptive innovation, helping us to reimagine transition not as a directional process of disruption encapsulated in the mud example, but as a multifaceted, plural process where directions may be accidental. Based on the literature on restorative and reparative justice, the following principles can help to identify forms of reparative innovation (Castán Broto et al., 2021):

- They recognize the wide range of impacts that climate change has on people's lives, from direct effects to psychological and cultural effects.
- They amplify and broadcast the voices of those who suffer most from climate change impacts and climate change action.
- They mobilize for forms of compensation, for example, through co-benefits.
- They create new social relationships between those who bear the brunt of the responsibility for carbon emissions and those who suffer its impacts.

However, this is not a call to generate and reproduce reparative innovation: reparative innovation is already happening, shaping the lives of people in an unwelcoming world. The bottom line is not only whether a reparative engagement with innovation is possible, but also whether such reparative imaginations will cast the transition as a response to the trauma of climate change. The reparative links to history, in this case, the environmental histories that have shaped the drivers of climate vulnerability and the infrastructure histories that have locked in carbon emissions in contemporary societies.

Where to fly to save ourselves in the middle of the storm is the plight of the bird in Bierstadt's painting. In her essay on the painting, Zhong Mengual asks us to find an alternative point of view to observe the storm unfolding in front of our eyes. She writes: "The imminence of the storm contains all of the immemorial moments of the dynamics of erosion" (2020, p. 251).

It is a painting of the history of that landscape, not as could be read in its geology, but as reflected in minute ephemeral moments in which multiple simultaneous actions conform to the landscape. The point of view of the mountain may embrace its passivity through the ages as much as the micro-moments in which change takes place. However, Zhong Mengual also calls the storm a "slow, fine sculptor" shaping the mountain in tiny strokes. Miller has argued that Bierstadt used the landscape aesthetic as a means to "escape history." His paintings cannot be understood as anything else than a tool at the service of moral, religious and national meanings: a mythic world sealed off from history (Miller, 2001).

As Zhong Mengual (2020, p. 250) observes, everything in Bierstadt's painting is "hurtling down" at great speed. They are specks of dust in the painter's brush. Only the bird flies in the opposite direction. It is a temptation to see every little reparative action as a futile effort in the eye of the storm. Yet, this painting offers little more than a fantasy through colonizing eyes. Not everything runs down in an organized, aesthetically pleasing way. The nature of change is chaotic, hence the fundamental fear it inspires. The question is whether a fragmented landscape of climate action will add up to a common direction of travel, as the bubbles in the mud of Rotorua seem to be doing. I like the bird's gesture because it matters in itself rather than because it has a guaranteed synergy with everything else in the mountain.

There is evidence of "reparative" forms of climate action. The reparative element is not incompatible with existing attempts at delivering social innovation, where there is an explicit acknowledgment of the historical legacy of exclusion and oppression. There is a little—perhaps even a lot of—hope in the possibility that we can mend the climate trauma. We hope for the time when hope no longer serves us. In the meantime, even brave little birds will shape the mountain. Minuscule actions sometimes translate into social change. Sometimes the time is ripe for change. Yet, no action is too small to catalyze change. Like the brave little bird, action's imprint on the landscape cannot be entirely discounted.

References

Acuto, M. (2014). Everyday international relations: Garbage, grand designs, and mundane matters. *International Political Sociology, 8*(4), 345–362.

Bulkeley, H. A., Castán Broto, V. & Edwards, G. A. (2014). *An urban politics of climate change: Experimentation and the governing of socio-technical transitions.* Routledge.

Carbon Brief Staff. (2021, November 15). COP26: Key outcomes agreed at the UN Climate Talks in Glasgow. Carbon Brief. https://www.carbonbrief.org/cop26-key-outcomes-agreed-at-the-un-climate-talks-in-glasgow

Carvalho, L., & Lazerini, I. (2018). Anchoring and mobility of local energy concepts: The case of community choice aggregation. In B. Turnheim, P. Kivimaa, & F. Berkhout (Eds.), *Innovating climate governance: Moving beyond experiments* (pp. 49–68). Cambridge University Press.

Castán Broto, V., & Bulkeley, H. (2018). Realigning circulations: How urban climate change experiments gain traction. In B. Turnheim, P. Kivimaa, & F. Berkhout (Eds.), *Innovating climate governance: Moving beyond experiments* (pp. 69–84). Cambridge University Press.

Castán Broto, V., Westman, L., & Huang, P. (2021). Reparative innovation for urban climate adaptation. *Journal of the British Academy, 9*(s9), 205–218. https://doi.org/10.5871/jba/009s9.205

Dave, N. N. (2017). Something, everything, nothing; or, Cows, dogs, and maggots. *Social Text,* 35(1), 37–57.

Eriksen, S., et al. (2021). Adaptation interventions and their effect on vulnerability in developing countries: Help, hindrance or irrelevance? *World Development, 141*, 105383.

Hanson, E. (2011). The future's eve: Reparative reading after Sedgwick. *South Atlantic Quarterly, 110*(1), 101–119.

Heiskanen, E., & Matschoss, K. (2018). Evaluating climate governance experiments: Participants' perspectives on low-carbon experiments in Finland. In B. Turnheim, P. Kivimaa, & F. Berkhout (Eds.), *Innovating climate governance: Moving beyond experiments* (pp. 182–200). Cambridge University Press.

Kareem, B., & Lwasa, S. (2011). From dependency to interdependencies: The emergence of a socially rooted but commercial waste sector in Kampala City, Uganda. *African Journal of Environmental Science and Technology, 5*(2), 136–142.

King, A. A., & Baatartogtokh, B. (2015). How useful is the theory of disruptive innovation? *MIT Sloan Management Review, 57*(1), 77–90.

Kivimaa, P., Laakso, S., Lonkila, A., & Kaljonen, M. (2021). Moving beyond disruptive innovation: A review of disruption in sustainability transitions. *Environmental Innovation and Societal Transitions, 38*, 110–126.

Kumar, A., Höffken, J. I., & Pols, A. (Eds.). (2021). *Dilemmas of energy transitions in the Global South: Balancing urgency and justice*. Routledge.

Latour, B. (2017). *Facing Gaia: Eight lectures on the new climatic regime*. Polity Press.

Latour, B., & Weibel, P. (2020). *Critical zones: The science and politics of landing on Earth*. MIT Press.

Liboiron, M. (2016). Redefining pollution and action: The matter of plastics. *Journal of Material Culture, 21*(1), 87–110.

L'Œil de la Photographie. (2019). Sophie Ristelhueber—Sunset years. https://loeildelaphotographie.com/en/sophie-ristelhueber-sunset-years-en/

Lwasa, S. (2019, September 2). Uganda offers lessons in tapping the power of solid waste. *The Conversation* https://theconversation.com/uganda-offers-lessons-in-tapping-the-power-of-solid-waste-119728

Lwasa, S., Kisembo, T., Mbabazi J., Namuli, H., Johnson, C., & Castán Broto, V. (2020). Integration of small-scale waste businesses into the urban economy. KNOW Policy Brief #1. Urban Action Lab, Makerere University. https://www.urban-know.com/policy-brief-1

Mbembe, A. (2019). *Necropolitics*. Duke University Press.

Miller, A. (2001). Albert Bierstadt, landscape aesthetics, and the meanings of the West in the Civil War era. *Art Institute of Chicago Museum Studies, 27*(1), 40–102.

Pallet, H. (2018). Understanding public dialogue as an embedded democratic innovation in UK climate governance. In B. Turnheim, P. Kivimaa, & F. Berkhout (Eds.), *Innovating climate governance: Moving beyond experiments* (pp. 85–102). Cambridge University Press.

Schipper, E. L. F., Eriksen, S. E., Fernandez Carril, L. R., Glavovic, B. C., & Shawoo, Z. (2021). Turbulent transformation: Abrupt societal disruption and climate resilient development. *Climate and Development, 13*(6), 467–474.

Secretariat of the UN Framework Convention of Climate Change. (2021). Nationally determined contributions under the Paris Agreement: Revised synthesis report by the Secretariat. Conference of the Parties serving as the meeting of the Parties to the Paris Agreement. Third session, Glasgow, October 31 to November 12, 2021. FCCC/PA/CMA/2021/8/Rev.1.

Sedgwick, E. K. (1997). Paranoid reading and reparative reading; or, You're so paranoid, you probably think this introduction is about you. In E. K. Sedgwick (Ed.), *Novel gazing: Queer readings in fiction* (pp. 1–38). Duke University Press.

Turnheim, B., Kivimaa, P., & Berkhout, F. (2018). Experiments and beyond. In B. Turnheim, P. Kivimaa, & F. Berkhout (Eds.), *Innovating climate governance: Moving beyond experiments* (pp. 1–26). Cambridge University Press.

Walker, M. U. (2010). *What is reparative justice?* The Aquinas Lecture, Marquette University.

Whitechapel Gallery. (2017). Sophie Ristelhueber—In conversation with David Company https://www.whitechapelgallery.org/events/sophie-ristelhueber-talk/

Wiegman, R. (2014). The times we're in: Queer feminist criticism and the reparative "turn." *Feminist Theory, 15*(1), 4–25.

Zhong Mengual, E. (2020). The point of view of the mountain. In B. Latour & P. Weibel (Eds.), *Critical zones: The science and politics of landing on Earth*. MIT Press.

About the author

Vanesa Castán Broto is Professor of Climate Urbanism at the Urban Institute, University of Sheffield. She leads the ERC project Low Carbon Action in Ordinary Cities (LoAct) and the GCRF project Community Energy and Sustainable Energy Transitions (CESET). She is the author of *Urban Energy Landscapes* (Cambridge University Press, 2019) and *Urban Sustainability and Justice* (with Linda Westman, Bloomsbury, 2020) and coeditor of the volumes *Climate Urbanism* (Palgrave, 2021) and *Inclusive Urban Development in the Global South* (Routledge, 2021).

3. Struggles for democratic decarbonization

Lessons from New York City

Ashley Dawson

Abstract
Moving from an account of the winter of 2021 in Texas (USA) and its impact on the electrical grid, Dawson details how corporate utilities, the primary providers of electricity to the majority of people in the US, are constitutively unprepared to cope with extreme weather. The essay then profiles the creative forms of direct action embraced by popular movements in New York against the expansion of fossil fuel infrastructure. In particular, it discusses the work of Public Power NY, a campaign that aims to get New York City off fossil fuels and move it to a power system based on democratically controlled renewable energy, offering a blueprint for what could be achieved on a bigger scale through a federal Green New Deal.

Keywords: social movements, direct action, community power, New York City

During the winter of 2021, my parents, who live in Austin, Texas, were forced to go without electricity and heat for 72 hours during a historic polar vortex. Power had been highly intermittent for days before the electric grid crashed. My parents, who are in their 80s, have no alternative source of heat in their house and could not get their car out of the driveway because of the ice and snow. Even if they had been able to leave home, where would they have gone? Austin only has one major warming center, miles away from where my parents live, and they were reluctant to go anywhere near large groups of people because of the pandemic. To make matters worse, they were also warned that their water had been contaminated.

Armiero, M., E. Turhan & S. Paolo De Rosa (eds), *Urban Movements and Climate Change: Loss, Damage and Radical Adaptation*. Taylor & Francis Group 2023
DOI: 10.5117/9789463726665_CH03

I followed the news of the climate/energy emergency in Texas from my home in New York City in utter horror for days, feeling totally helpless. I was thankful each morning when I received a text from my parents reassuring me that they were still alive. Fortunately, they had put gas in their car before the storm hit: after their home went dead, the car became their warming and phone-charging station. But this was slim solace: What would happen if their gas ran out, they wondered, during one of our brief phone calls? It was a matter of grim survival for them and their neighbors for several harrowing days.

The climate/energy emergency in Texas was no natural disaster. Many scientists argue that the extreme cold in Texas was generated by a polar vortex that pushed so far south as the result of an interruption to the jet stream caused by climate change.[1] The fragility of the grid in Texas was also a result of miscalculation, and the folly and greed produced by hyper-capitalism. Power authorities in the state typically expect high energy demand during the region's blistering summers, but relatively low demand in the winter. Power was down across the state because people had been consuming more energy to warm their homes, leading to demand levels that far outstripped supply. The Texas power authority, known as the Electricity Reliability Council of Texas (ERCOT), did attempt to prepare for the cold snap, but its models were based on previous weather patterns,[2] and the climate emergency means that the past is no longer a good model for the extreme future.

To make matters worse, since Texas operates its own independent electricity grid, it could not tap power supplies from neighboring states where demand was not so high. The state has a long history of allowing its private utilities to operate without regulation (Galbraith, 2021). Even after the formation of the state's official utility regulatory body, ERCOT, in the 1970s, utilities had little public supervision since its board was stocked with electric-industry insiders. It should come as little surprise, then, that the state's utilities channeled profits to investors and corporate CEOs rather than to building backup systems that could have saved millions of Texans from huddling in the freezing cold over that week.

Texas's energy crisis may have been particularly grave, but it is far from an isolated case. Today's electric grids in the United States are a relic of

1 The jet stream has historically trapped cold air around the North Pole. According to scientists, climate change is interrupting the air currents that form the jet stream, which, in turn, is allowing frigid air to sweep south into continental North America. See Milman, 2021.

2 "Seasonal Assessment of Resource Adequacy for the ERCOT Region," http://www.ercot.com/content/wcm/lists/197378/SARA-FinalWinter2020-2021.pdf.

the early 20th century. They rely on centralized generation of power that is distributed through a frail web of aerial arteries to distant, isolated homes. These systems were built to withstand freak accidents, yet, as physicist and energy activist Amory Lovins explained decades ago, "possible rare events, each of vanishingly low probability, are infinitely numerous, so we live in a world full of nasty surprises" (Lovins, 1977). As prescient as this warning about grid fragility was, even Lovins could not have anticipated how these nasty surprises would proliferate in the age of the climate emergency. Not half a year after the climate/energy emergency in Texas, for example, a heat wave in the Pacific Northwest drove temperatures above 100°F (37.7°C), melting the power cables for streetcars in the city of Portland, Oregon, and leading to power outages for thousands of people. Within days of this emergency, officials in New York City sent out an emergency mobile alert urging residents to conserve energy by cutting electricity use during a devastating heat wave.

The grid is under unprecedented pressure as a result of the climate emergency. In this chapter, I show that corporate utilities, the primary providers of electricity to the majority of people living in the US, are constitutively unprepared to cope with extreme weather conditions. Not only have they placed private profit over public safety for decades, but they themselves have helped generate the climate emergency that is now stressing the grid to the point of collapse by fighting against meaningful steps toward a transition away from polluting and planet-heating fossil fuels. This chapter profiles the creative forms of direct action embraced by popular movements in New York against the expansion of fossil fuel infrastructure. These movements oppose the city's corporate utilities, which are pushing forward fossil infrastructure over the objections of climate scientists. In addition, this chapter discusses the work of Public Power NY, a campaign that emerged from the realization that private power utilities are killing us and the planet with their greenhouse gas-spewing projects. The Public Power NY campaign aims to get the nation's biggest city and one of its most progressive states off fossil fuels and move it to a power system based on democratically controlled renewable energy. The campaign has been seen as offering a blueprint for what could be achieved on a bigger scale through a federal Green New Deal to provide clean energy for the nation (Noor, 2021).

I have written at length about the energy commons and democratic decarbonization in my book, *People's Power* (Dawson, 2020). A brief recapitulation of some of my key arguments here will clarify, I hope, some of the stakes in the struggle taking place in New York, and it will also illuminate the implications of this local struggle for broader fights against fossil capitalism. Historians such as Timothy Mitchell have offered detailed analyses of how elites used the material characteristics of petroleum during the mid-20th century to defeat working-class movements: oil, Mitchell argues, is relatively easy to extract and transport in comparison with coal, and therefore was embraced by elites in core imperial nations as a means to break worker power organized around control of energy and transport infrastructures. Building on Mitchell's attention to the materiality of energy and attendant infrastructures, I point out in *People's Power* that, in comparison with fossil fuels, forms of renewable energy such as wind and solar power are widely distributed and easily accessible. They therefore hold the potential for popular empowerment, in both senses of that term, a potential I argue is powerfully symbolized by the idea of energy as a commons, a collectively stewarded resource accessible to all. Activists such as Denise Fairchild and Al Weinrub (2017) have argued for community-based renewable energy cooperatives as the embodiment of this popular empowerment. Such forms of energy governance are undeniably a welcome antithesis to kleptocratic corporate utilities, but the climate emergency dictates that energy transition must be swift. Activists therefore need to think about how to scale up the energy commons, while maintaining forms of democratic and egalitarian governance for the commons. This is the challenge embraced by the Public Power NY campaign in New York.

The refusal of New York state legislators to take meaningful steps toward energy transition has ominous implications not just for New Yorkers, but for the country and the planet: electricity production currently generates a quarter of US carbon emissions, but since we need to electrify everything—including carbon-intensive sectors like transportation and the heating/cooling of homes and buildings—this is a decisive fight in the struggle against climate breakdown (Roberts, 2017). How, this chapter asks, do movements for energy commoning win what Alyssa Battistoni and Jedediah Britton-Purdy (2020) call democratic decarbonization? To the extent that ideas about public power as a solution to the increasingly apparent failures of investor-owned utilities have become part of public discourse in the course of just a few years, it could be said that we've already turned the tables on fossil capitalism. This is no small thing since the majority of US residents have gotten their power from investor-owned utilities since the creation of the modern grid over a century ago. The fact that the idea of public power is being taken on board by increasing numbers of elected officials adds to a sense of gathering momentum. But how do energy commons activists ensure that legislation mandating democratically governed renewable energy is passed? And how do we do it in the short amount of time we have before the climate emergency intensifies unstoppably?

In analyzing what I call Blockadia NYC and the Public Power NY campaign, I emphasize the dual character of political movements against fossil capitalism: they hinge both on defending people's lives and communities against the forms of environmental injustice carried out by investor-owned energy companies and the state, and also on moral condemnation of the society that permits such repugnant injustices to take place. Struggles over resources are thus always simultaneously struggles over meaning, as the environmental historians Ramachandra Guha and Joan Martinez-Alier (1997, p. 13) have argued in their analysis of environmental movements in India. To extend their argument, I suggest this means that environmental struggles tend to display a strong cultural politics, with dissident viewpoints articulated through potent aesthetic forms of expression. I use articulation here in the specific sense used by cultural studies theorists such as the Black British intellectual Stuart Hall, where the term is used to designate the simultaneous effort to convey meaning and to forge connection (Hall et al., 1996, p. 115). Articulation is thus simultaneously an expressive tool and a weapon of political organizing. In this way, aesthetic expression and the cultural politics that attend resistance to fossil capitalism more broadly are characterized by powerful expressions of dissent.

Articulation also takes place across spatial scales. As my research demonstrates, political struggles that unfold on the urban terrain of a city like New York become staging grounds for regional and ultimately national campaigns for energy democracy. Moreover, political mobilizations against fossil capitalism such as those that unfolded in New York in recent years have a notably intersectional quality: the construction of a pipeline through a predominantly Black or Latinx community is seen by organizers, for example, as directly related to systemic injustices that generate gentrification, violent policing of communities of color, the prison—industrial complex, and racial capitalism and settler colonialism more broadly. This means that mobilizing people against fossil capitalism and for public power involves establishing links between their own specific experiences and broader forms of oppression. Beyond this, it also means establishing chains of equivalent oppressions around the disparate experiences of populations subjected to racial capitalism and settler colonialism. Here again the term "articulation" is useful since it helps explain the ways in which activists challenge and reorient dominant perspectives on fossil capitalism, simultaneously forging potent oppositional coalitions fighting to reclaim the energy commons and public power more broadly. This is how a world of solidarity beyond fossil capitalism will be forged.

Power failures

Fossil capitalism is hacking away at the teetering foundations of society in the US. Recent examples of climate change-related disasters are legion. Meanwhile, in late June, an ExxonMobil lobbyist was caught in a Greenpeace sting video describing his firm's strategy to gut meaningful climate provisions from the Biden administration's infrastructure bill (Tabuchi, 2021). The cynicism of the corporation's tactics was breathtaking: the lobbyist described his firm's public support of a tax on carbon emissions as a "great talking point" for the company, which believes that such a tax will never become law. Of course, Exxon's ecocidal behavior is not exactly news: fossil fuel companies have a long history of actively broadcasting disinformation about the climate crisis in order to quell efforts to shift away from consumption of fossil fuels (Hall, 2015).

The investor-owned utilities that provide most Americans with electricity engage in similar dirty political tricks in order to maintain their economic stranglehold over communities. For example, utilities around the country have done their best to slow the growth of renewable energy by supporting laws and regulations that reduce targets for renewable energy. These kinds of anti-renewables measures are often generated by the right-wing American Legislative Exchange Council (ALEC). Legislators use ALEC boilerplate language to write anti-renewables laws and take campaign donations from industry groups like the Edison Electric Institute, which in turn funds ALEC (Surgey, 2014).

All of this blatant corruption has left the US with a shambolic grid that is straining, and, in some cases, buckling under the stress of increasingly extreme weather. From the combined energy and political crisis during winter 2021 in Texas to blackouts across the Pacific Northwest in the midst of a dangerous heat wave this past summer, climate change is generating punishing conditions for which the nation's creaking energy infrastructure is ill equipped.

Meanwhile, citizens are becoming increasingly dependent on reliable power for access to life-saving cooling and heating during the extreme climatic conditions that are laying bare the frayed state of the electric grid. It is a savage irony that this country's poorest people, those who contribute the least to carbon emissions, are most at risk when the power goes out, leaving them exposed to heat- and cold-related illness and death (Chabria & Luna, 2015). The stress on the grid and on vulnerable communities is only going to build as climate change produces more intense weather extremes and as political pressures to dump fossil fuels intensify.

The collapse of the grid in Texas was, according to Naomi Klein, the terminal failure of a 40-year experiment in free-market fundamentalism (Klein, 2021). Yet if New York has not suffered such a cataclysmic failure, it is equally subject to the baleful influence of powerful private energy companies. Like those in Texas, these companies put returning profits to shareholders and CEOs above maintaining the grid, meaning that power outages are an increasingly frequent occurrence (Zaveri & Wong, 2021). While it could be argued that this is the mission of all capitalist corporations, the neoliberal deregulation of the Texas energy grid generated a particularly disastrous situation in which the grid was perpetually on the edge of collapse.

Worse still, New York's power mix is very dirty: despite the state's much-ballyhooed Reforming the Energy Vision (REV) initiative, only 6% of New York City's energy currently comes from renewable sources like solar and wind (New York Independent System Operator, 2019). America's largest city is actually slated to burn far more fossil fuels in the future because the nuclear power plant that supplies 80% of its nonfossil energy shut down in April (US Energy Information Administration, 2021). Given the failure of investor-owned utilities to build out renewable power, the deficit in energy produced by this closure is going to come from fossil fuel-burning "peaker" plants. Most of these dirty plants are located in communities of color, leaving their inhabitants to pay with their lungs and their lives for the energy consumed by their fellow citizens to cope with extreme weather.

Blockadia NYC

On a sunny day in early May, I joined a flotilla of bicycles, skateboards and even a disco dancer on roller skates, rolling through the streets of Brooklyn. With a loudspeaker on the back of a cargo bike pumping out tunes, our parade felt joyful. But we had gathered for a sober purpose: to protest a fracked-gas pipeline being constructed through predominantly Black and Brown working-class neighborhoods in North Brooklyn by the corporate utility National Grid. As we cycled the pipeline route, we handed out a 'zine drawn by local artists to onlookers, explaining that the pipeline is part of National Grid's planned expansion of a liquified gas depot on the nearby Newtown Creek. This Superfund site is polluted by a century and a half of fossil fuel dumping. Now, community members are being stuck with nearly half a billion dollars in rate hikes to pay for yet more polluting infrastructure that will keep the city locked into fossil fuels for decades.

What is a fracked gas pipeline doing in NYC? Since the beginning of the fracking boom nearly two decades ago, the US has been on the highway to climate hell, despite the repeated warnings of scientists that further drilling for fresh reserves of gas and oil would push the climate toward dangerous tipping points. Notwithstanding these warnings, the myth that "natural" gas was a viable "bridge" fuel to take us from an era of petroleum to the age of clean energy took hold, leading both the Obama and Bush administrations to support fracking with billions of dollars in subsidies. Trump promised to make things far worse by bringing back coal, but largely stuck to the policies of predecessor administrations. New York seemed to be on a different path after grassroots movements successfully fought to ban fracking in the state. Although fracked gas is not actually being brought to the surface in New York, that does not mean that we don't have any gas infrastructure: utility companies are still importing as much fracked gas as possible from nearby centers of production like Pennsylvania.

The climate movement has shut down major gas infrastructure projects like the Port Ambrose Liquified Natural Gas (LNG) terminal and the Williams NESE pipeline, but through utility bills rate payers in the state have been forced to expand and extend the life of small, more distributed fracked gas infrastructure.[3] Even ostensibly "green" political leaders in the state like Mayors Bloomberg and De Blasio and Governor Cuomo have largely supported this expansion of fossil gas infrastructure, often while simultaneously mouthing support for trend-setting environmental legislation like the Climate Leadership and Community Protection Act (CLCPA) advanced by frontline environmental justice groups. While such ambitious climate legislation pledges significant reductions in greenhouse gas emissions and expansion of renewable energy, the promised reductions were a decade or more away, allowing political leaders to talk out of both sides of their mouths at once. Even after Cuomo signed the CLCPA into law, for example, the New York Public Service Commission (PSC), the state's energy regulatory body, approved the request of a corporate utility, Consolidated Edison (popularly known as Con Ed), to raise rates in downstate New York to spend almost a billion dollars on new fracked gas infrastructure.[4]

"Enough is enough," members of predominantly Black and Brown communities in Brooklyn like Gabriel Jamieson said when they learned of National Grid's plans to build a pipeline and liquified natural gas terminal in their community. Jamieson had worked for several years to get hundreds

3 Lee Ziesche, personal communication (December 4, 2020).

4 Ibid.

of his neighbors in Brownsville to install rooftop solar panels through a program called Solar Pioneers. In fall 2019, Jamieson and seven other community members climbed down into the trenches dug by National Grid and chained themselves to the pipeline to protest against the company's plans. Community members like Jamieson have been battling for decades against environmental racism, for the cleanup of a massive oil spill in North Brooklyn neighborhoods like Greenpoint, against toxic particulate matter from the major highway that slices through the area, and against the pollution caused by trucks idling at waste transfer stations in Williamsburg and Bushwick (see Willyard, 2012).

Communities in Jamieson's neighborhood of Brownsville, where the pipeline begins, have the highest rates of asthma in the city. Instead of more dangerous fossil infrastructure, protesters with Frack Outta Brooklyn and allied organizations joined the statewide Public Power NY coalition in demanding clean power, as well as political power over their utilities, communities, health and futures.[5] As activist Lee Ziesche put it, "We're at a crossroad: we can follow the utilities down a path where their shareholders continue to make obscene profits while communities shoulder the costs, or we can follow the lead of the Black and Brown communities who are standing up to National Grid's racist fracked gas pipeline and demanding to take back their power."

The 'zine we distributed to onlookers as we rode with Frack Outta Brooklyn and allied organizations proclaimed, "We Are Not a Sacrifice Zone" (see Figure 3.1). The language of the "sacrifice zone" connects the struggle in North Brooklyn to a long history of environmental injustice in the US that includes the use of Native American lands for nuclear testing, strip mining for coal in working-class rural areas and toxic exposure to chemicals produced by fossil fuel-related industries such as has occurred in along the lower stretch of the Mississippi River in Louisiana, an area dubbed "cancer alley" for the high incidence of illness among its predominantly Black residents. In this way, protest organizers linked the struggle of North Brooklyn residents to fights against environmental racism that have been a core element of the environmental justice movement over the last half-century.

The image that accompanied this proclamation also made multiple connections. In this image, a circular pipeline surrounds a black trifoil, the international symbol for hazardous radioactive material. Juxtaposed over the trifoil is a young Black girl holding a protest sign against the Brooklyn pipeline. The image is shocking, recalling the public terror that circulated

5 Public Power NY, https://ecosocialists.nyc/public-power/.

Figure 3.1. The cover of the 'zine *We Are Not a Sacrifice Zone*, Brooklyn, NY. Artist: Andrea Lomanto.

during the Three Mile Island partial nuclear meltdown in 1979. There is a generalized rhetoric of toxicity that the image summons, conjuring up the long history of racist subjection of communities of color to toxicity. But the image also alerts local residents to specific linkages between fossil gas extraction and exposure to hazardous radioactive material. Although the gas industry has tried to keep this information secret, it turns out that radon and other radioactive elements build up inside pipelines bringing fracked gas from shale formations in Pennsylvania to New York, endangering utility workers and local residents alike (Nobel, 2020).

The images mobilized by the campaign against the pipeline also make connections with other struggles against fossil infrastructure in the US. For example, a poster distributed during the COVID-19 pandemic features a black snake whose coils are composed of pipeline sections, each of which is labeled with a different Brooklyn neighborhood through which

Figure 3.2. An anti-pipeline poster from the No NBK Pipeline movement. Artist: Kim Fraczek.

the National Grid project passes (see Figure 3.2). The image of the black snake derives from the Lakota prophecy that became central to Native American resistance to the Keystone XL pipeline in North Dakota in 2016. This baleful image has been mobilized once again by Native Americans and allies fighting against the Enbridge Line 3 pipeline in Minnesota today. By deploying the snake image, in other words, campaigners in Brooklyn not only proclaimed their solidarity with protesters in the Dakotas and in Minnesota, but also articulated their struggle as inextricably intertwined with those fights. The image helps protesters and the broader general public see fossil capitalism as a system, so that local struggles against specific pipelines and other elements of infrastructure can be seen as part of a broader fight against an industry that is driving the planet toward extinction.

What's more, fossil capitalism is linked to other death-dealing aspects of a system that hinges on white supremacy. For this reason, campaigners made connections between environmental injustice and other forms of structural violence against people of color in the present. "Climate Justice = Racial Justice. Black Lives Matter," activists against the North Brooklyn pipeline proclaimed. The forms of violence to which communities are subjected by fossil capitalism are, in other words, linked to the violence exerted against them by a militarized police force, and by the racial capitalist system more broadly. In this way, the struggle against the North Brooklyn pipeline is also a struggle against a system that, while driving the entire planet toward ecocide, has clearly marked out certain populations for accelerated extermination. The fight against the black snake gives the lie to the idea that cheap and abundant fossil fuels are the basis of unbridled freedom and ceaseless growth in the world's developed nations.

Public power

The struggle for democratic control over energy production, distribution and use is a key front in the fight for climate justice. As long as energy production remains grounded in the logic of capital accumulation, it will continue to obey the irrational drive to generate ceaselessly expanding profits off fossil fuels and damn the environmental and social consequences. Since fossil fuel corporations have immense amounts of assets sunk into existing infrastructures, they will fight tooth and nail to prevent a transition to a low-carbon society, despite the occasional charade of moving "beyond petroleum." Truly sustainable energy production will only be possible if power is taken out of the hands of capital and the capitalist state. Power, in both senses of the term, must consequently be controlled by ordinary workers and communities.

The Public Power NY campaign emerges from the realization that corporate monopolies like National Grid and Con Ed are killing us and the planet with their greenhouse gas-spewing projects. These for-profit corporations are owned and controlled by Wall Street investors. Instead of serving the public interest, they make billions of dollars in profits each year for their wealthy shareholders by charging the nation's second-highest rates for services that are unreliable and often hazardous. Begun only two years ago in response to a proposed rate hike for consumers in New York City, the Public Power NY campaign is an ambitious effort to rebuild the state's electric grid around democratically controlled renewable energy. Over the last year, the campaign created two game-changing bills that gained dozens of cosponsors within the New York state legislature: the New York Build Public Renewables Act (NYBPRA) and the New York Utility Democracy Act (NYUDA).

The first of these bills proposes to expand the New Deal-era New York Power Authority (NYPA), which currently provides reliable and affordable power to public entities like the CUNY and SUNY campuses and to public housing across the state. Under NYBPRA, the New York Power Authority would be empowered to own and build new renewable generation, storage and transmission facilities, with an initial aim of providing 100% renewable energy to all state and municipal properties and transportation by 2025. The bill also directs NYPA to sell renewable power directly to consumers, providing them with an alternative to for-profit utilities. There has been much debate within the movement about what exactly constitutes socially just renewable energy, particularly since New York City is currently pursuing plans to draw "clean" power from dams in Quebec, a scheme that has been denounced by environmentalists and indigenous activists (Amarnath et al., 2019). Although somewhat beyond the scope of NYBPRA, the Public Power NY campaign intends the focus of the new authority to be on "modern" renewables such as wind, solar and geothermal power.

Convincing labor unions that support for fossil capitalism damages their interests is a key part of the Public Power NY campaign. The energy industry has long pursued a strategy of pitting labor unions against environmentalists, an approach that has garnered support from building trades unions for fossil infrastructure like the Keystone XL pipeline.[6] The fracking boom has undeniably generated many well-paying jobs, but for-profit utilities have a history of trying to bust their workers' unions (Wilbur, 2012). In addition, the energy industry isn't just poisoning the communities near wells, they're also poisoning their workers. A recent investigation revealed that pipelines are coated with radioactive elements like radon that give workers sores, skin lesions and cancer (Noble, 2020).

Workers deserve a better alternative. To that end, all projects under the newly reconstituted NYPA would pay prevailing wages and be subject to project labor agreements, ensuring a just transition for people currently working for corporate utilities. According to a recent report by the Community and Climate Project and the Democracy Collaborative, NYBRPA

6 "NAM, LIUNA, NABTU Applaud Keystone XL Pipeline Permit" (March 29, 2019), https://nabtu.org/press_releases/nam-liuna-nabtu-applaud-keystone-xl-pipeline-permit/.

would create between 28,000 and 51,000 well-paying jobs.[7] The expansion of renewable energy infrastructure mandated by the Public Power NY legislation would add an estimated $90 billion to the state's economy at a time of dire need for working-class New Yorkers.

The Public Power NY campaign is also interested in bringing genuine energy democracy to a sector that has long been marked by the capture of public regulatory bodies by oligarchical corporations. State regulators have been caught meeting behind closed doors with utility industry lobbyists in states like Texas, leading to nullification of the Clean Air Act regulations for coal plants. The campaign's second element, the NY Utility Democracy Act (NYUDA), would initiate a transition of the state's distribution utilities from private to public ownership within two years. Democratically elected utility boards would be created to oversee the operations of the new public distribution utilities. The measure would also order the creation of an observatory oversight body with representation from environmental justice organizations, labor, community groups and engineering experts. This reorganization would help make real the promise of energy as a commons, a collectively owned and controlled public good.

If popular discontent with neoliberal globalization is not to be even more thoroughly captured by the forces of reaction, we urgently need to articulate viable models of public ownership based on principles of economic egalitarianism, decentered decision-making and public participation that reanimate the idea of democracy and the common good (Cumbers, 2016–2017, p. 84). We need new narratives of public potential that inspire a sense of radical possibility and progressive popular struggle. In no realm is this reassertion of public prerogatives more important than in the struggle for energy democracy. Our collective survival literally hinges on this fight. The current centralized model of energy, a legacy of the age of fossil fuels, is characterized in the US by the control of energy generation and distribution by a small number of large, for-profit corporations. Concentrating power and wealth in the hands of a few, this centralized form of power is the model of choice for corporate renewable energy transition (Fairchild & Weinrub, 2017, p. 12). But renewable energy resources are inherently distributed. Rather than generating electricity from one central power plant, that is, renewable energy comes from solar power, wind and geothermal resources that can be found and harvested everywhere. The movement for energy democracy is thus fighting for decentralized forms of power in both senses of

7 "A New Era of Public Power" (April 22, 2021), https://democracycollaborative.org/a-new-era-of-public-power.

the term: energy that is generated locally, energy that takes power away from today's economic, political and energy oligarchies in order to empower the frontline communities—working people, indigenous groups, communities of color—who bear the brunt of fossil capitalism's toxic environmental, economic and repressive elements.

New York as laboratory

In the 1930s, New York was a laboratory for many policies central to the federal New Deal. The Public Power NY campaign is renewing the city and state's role as a bellwether of progressive change on a national level. In June 2021, nearly two years after the inception of the campaign in New York, Congressman Jamaal Bowman of New York and Congresswoman Cori Bush of Missouri introduced a public power resolution to Congress. The resolution is an index of the tremendous ideological work and transformation brought about by the campaign in New York. The one-pager distributed to publicize the legislation, for instance, begins with the forthright statement: "Time to Kick Out Corporate Utilities, Time for Energy Democracy." The very framing of the entities that for most of the last hundred years have been known as "public utilities" as corporations is notably radical. It's worth noting that the idea of "public utility" was a strategy adopted by the corporate utilities during the Progressive Era in order to quash municipalization campaigns that were fast normalizing public power (Dawson, 2020, pp. 72–95). During the New Deal era, when the corruption and flagrant price-gouging associated with the utilities became the object of public knowledge and opprobrium, a deal was struck to allow state regulation of corporate utilities through entities such as New York's Public Service Commission. In this way, the idea that the corporate utilities functioned in the public interest became ideological common sense. That idea is being dismantled by today's campaigns.

Bush and Bowman's public power resolution offers a litany of devastating arguments against corporate utilities. Drawing on many of the perspectives developed first by frontline environmental justice campaigns and then by the Public Power NY campaign in New York, the resolution cites the constitutively flawed character of the investor-owned utilities, which repeatedly place profits before the public good: "Our energy systems repeatedly exacerbate disasters we know are worsening and take lives in the process of prioritizing shareholder profits over people's needs.... To save lives and fight the climate crisis at the same time, utilities must be public goods for everyone."[8] The resolution stresses the ways in which corporate utilities are unable to prevent energy poverty, unwilling to prioritize updating their equipment over investor profits and unfit to cope with the climate crisis. As was true of the Public Power NY campaign in New York, the Bush and Bowman resolution places energy democracy in the context of intersectional struggles for a just, resilient world: "The United States must re-imagine its power system to be just, equitable, anti-racist, and climate- and disaster-resilient through establishing a public power system."[9]

As befits proposed legislation on the federal level, Bush and Bowman's resolution aims to transform and integrate the fragmented US energy system. Thus, it proposes to support existing public and cooperative energy provides to transition to 100% renewable energy. It also requires that existing large power providers such as the Tennessee Valley Authority, an organization established during the New Deal that is currently wedded to significant fossil fuel sources of energy generation, move speedily to genuinely clean energy. Like the Public Power NY campaign, it envisages a transition from investor-owned power to renewable public power, in this instance by "establishing aggressive targets for investor-owned utilities' decarbonization … and bringing noncompliant utilities into an appropriate scale of public ownership." Last of all, the proposed legislation fights energy poverty by calling for the enactment of a universal ban on electricity disconnections for nonpayment.

Bush and Bowman's public power resolution initially had seven cosponsors: progressive members of Congress like Alexandria Ocasio-Cortez who collectively have become known as the so-called Squad. It is worth noting that these legislators represent precisely the frontline communities impacted by the environmental injustices being fought by groups like Frack Outta Brooklyn. Cori Bush herself was priced out of her home by astronomical energy bills in the impoverished, predominantly Black city of Ferguson. It may be true that there are not many other members of Congress who can draw on such galvanizing experiences. Indeed, many in Congress have a long history of accepting campaign donations from corporate utilities and fossil fuel corporations. But the Public Power NY campaign has begun to change the conversation in Congress and within the Democratic Party. As a result of this campaign, and after disasters such as the one in Texas during

8 https://bowman.house.gov/_cache/files/1/7/1758fcd2-28b1-433a-a235-69af2b47dc30/66B21CF944634B42143E1AD716D6AE97.bush-bowman-public-power-one-pager.pdf.

9 https://bush.house.gov/sites/evo-subsites/bush.house.gov/files/evo-media-document/Bush Public Power Resolution FINAL.pdf.

the winter of 2020, the need for public power is becoming more and more apparent to the American public. Greater numbers of people now understand that corporate utilities are responsible for the climate crisis. It is becoming common sense that fossil capitalism has got to go.

The road forward

The public power legislation should have carried easily in New York state, which enjoys a democratic supermajority. With the profit motive removed from the energy sector, utility rates would have gone down. The two proposed bills would bring New York's energy infrastructure on target to meet the ambitious energy transition goals dictated by the Climate Leadership and Community Protection Act (CLCPA) of 2019. But instead of beginning to fulfill the CLCPA's promises, elected leaders punted the chance to clean up our electric grid and establish a nation-leading model for an equitable green recovery. An analysis of campaign finance data laid bare at least one important reason for this inaction: bribes to leading state politicians from for-profit utilities.[10]

Public Power NY advocates are currently assessing the work done over the last year and strategizing about how to build political power so that the campaign will ultimately be successful. A few things are already clear. According to Patrick Robbins, coordinator of the New York Energy Democracy Alliance, ensuring that the campaign and the bill support the labor movement is key.[11] So, Robbins says, is spreading the campaign throughout New York state since fossil capitalism has all too often tried to split coalitions along class lines by suggesting that people in New York City treat the rest of the state as a sacrifice zone.

In addition to this solidarity-building work, climate activists must make the price of constructing more world-destroying fossil infrastructure so steep that renewable alternatives become publicly accepted as the only way forward. Direct action protest is a key component of this struggle. As Lee Ziesche of the Sane Energy Project told me, National Grid only halted construction of its fracked gas pipeline in North Brooklyn after activists locked themselves to the pipeline to shut down work, thereby alerting the community and the media to the fact that a polluting and dangerous

10 "How Fossil Fuel Companies Bought off New York Democrats," https://ecosocialists.nyc/ny-climate-bribe/.

11 Patrick Robbins, personal communication (July 1, 2021).

project was being built.[12] The movement is currently leading community members in a Gas Bill Strike, refusing to pay monthly rates that fund the construction of dangerous fossil infrastructure.

Although movements curtailed public protests during the pandemic lockdown, and legislators across the country are increasingly seeking to outlaw civil disobedience, the struggle against fossil capital demands increasingly creative forms of public disruption (Sen, 2021). The Public Power NY campaign can draw on tactics developed during other struggles, from the civil rights movement to ACT UP. Politicians who take money from or otherwise support fossil capitalism need to be called out or zapped in public. The work of community education and empowerment against fossil capitalist infrastructure can be done through rabble-rousing, bicycle-led, boom-box-beat-laden toxic tours like the one organized by the No North Brooklyn Pipeline campaign this spring. Let a million Climate Strikes bloom! Activists may even turn to sabotage of fossil capitalist infrastructure (Malm, 2021). Such disruptive work must remain nonviolent if the movement is going to continue to win adherents, but given the obdurate determination of fossil capital to keep burning up the planet, fossil infrastructure cannot remain untouchable forever. As scholar-activist Andreas Malm puts it, respect for such destructive private property will cost us the earth (Malm, 2021, p. 68).

Now more than ever, energy democracy activists must adopt what has been called an "in-against-and-beyond" the state struggle (Angel, 2016). Contemporary movements like Public Power NY are trying to make big changes by working within the existing political system. Given the long tradition of leftist aversion to any engagement with state power, this can feel refreshing and even radical: instead of waiting for an outright revolution to smash the state, this generation of activists is proposing big structural changes and then pushing the establishment to embrace them or get out of the way. But over the last year the Public Power NY campaign has run headlong into the inertia and corruption of the political system even in a nominally progressive state like New York. It has consequently become clear that the campaign can only succeed by working with sympathetic legislators while simultaneously adopting the direct-action tactics of autonomous social movements. In this way, the Public Power NY campaign can use the gathering momentum of the climate insurgency against fossil capital to build pressure on legislators to support energy transition. In other words, democratic decarbonization requires people's power in the streets.

12 Ziesche, personal communication.

References

Amarnath, K., Dawson, A., & O'Reilly, S. (2019, May 20). Bill de Blasio's energy plan isn't as clean as it looks. *The Guardian*.

Angel, J. (2016). Towards an energy politics in-against-and-beyond the state: Berlin's struggle for energy democracy. *Antipode, 51* (4), 1365-1365. https://doi.org/10.1111/anti.12289.

Battistoni, A., & Britton-Purdy, J. (2020). After carbon democracy. *Dissent, 67*(1), 51–50.

Chabria, A., & Luna, T. (2015, October 11). PG&E power outage brings darkness, stress, and debt to California's poor and elderly. *Los Angeles Times*.

Cumbers, A. (2016–2017). Public ownership as economic democracy. *Soundings: A Journal of Politics and Culture*, 64, 84–88.

Dawson, A. (2020). *People's power: Reclaiming the energy commons.* O/R Books.

Fairchild, D., & Weinrub, A. (Eds.). (2017). *Energy democracy: Advancing equity in clean energy solutions*. Island Press.

Galbraith, K. (2021, February 16). There are 3 power grids in the lower 48 states, and Texas has its own. Here's why. *Fort Worth-Star Telegram*.

Guha, R., & Martinez-Alier, J. (1997). *Varieties of environmentalism: Essays north and south*. Earthscan.

Hall, S. (2015, October 26). Exxon knew about climate change almost 40 years ago. *Scientific American*.

Hall, S., Morley, D., & Chen, K. H. (1996). *Stuart Hall: Critical dialogues in cultural studies*. Routledge.

Klein, N. (2021, February 21). Why Texas Republicans fear the Green New Deal. *New York Times*.

Lerner, S., & Brown, P. (2012). *Sacrifice zones: Front lines of toxic chemical exposure in the United States*. MIT Press.

Lovins, A. (1977, July 24). Resilience in energy strategy. *New York Times*.

Maldonado, J. K. (2018). *Seeking Justice in an energy sacrifice zone: Standing on vanishing land in Louisiana*. Routledge.

Malm, A. (2021). *How to blow up a pipeline*. Verso.

Milman, O. (2021, February 17). Heating Arctic may be to blame for snowstorms in Texas, scientists argue. *The Guardian*.

New York Independent System Operator. (2019, July 17). Fuel for the wire: How we make energy in New York.

Nobel, J. (2020, January 21). America's radioactive secret. *Rolling Stone*.

Noor, D. (2021, May 13). New York's public power bill could be a model for the rest of the country. *Gizmodo*.

Roberts, R. (2017, October 27). The key to tackling climate change: Electrify everything. *Vox*.

Sen, S. (2021, July 2). Fossil fueled-crackdowns are part of the assault on democracy. *Newsweek*.

Surgey, N. (2014, July 30). Utility trade group funds ALEC attack on Americans using solar." *PR Watch*. https://www.prwatch.org/news/2014/07/12553/utility-trade-group-funds-alec-attack-americans-who-use-solar

Tabuchi, H. (2021, June 30). In video, Exxon lobbyist describes efforts to undercut climate action. *New York Times*.

US Energy Information Administration. (2021, April 30). Today in energy: New York's Indian Point nuclear power plant closes after 59 years of operation. *Today in Energy*. https://www.eia.gov/todayinenergy/detail.php?id=47776

Wilbur, M. (2012, July 16). Locked out Con Ed workers: Utility trying to bust union. *The Examiner*.

Willyard, C. (2012, January 5). New York's dirty secret: The effort to clean up America's largest oil spill. *Earth Magazine*.

Zaveri, M., & Wong, A. (2021, June 30). "Conserve energy": New York City begs residents to help avoid outages." *New York Times*.

About the author

Ashley Dawson is Professor of Postcolonial Studies in the English Department at the Graduate Center, City University of New York and the College of Staten Island. He is the author of numerous books, including the freshly published *People's Power: Reclaiming the Energy Commons* (O/R, 2020). A member of the Social Text Collective and the founder of the CUNY Climate Action Lab, he is a long-time climate justice activist.

4. Disobey, block, organize

The politics and strategies of grassroots climate activism in Malmö and Sweden

Salvatore Paolo De Rosa

Abstract

This chapter moves from a characterization of top-down, technocratic and market-oriented interventions in climate governance by state and corporations as hegemonic forces of depoliticization. Such a consensual regime of climate change governance leaves untouched the power of capital, the rule of the market and the global imperial hierarchies underpinning both. The chapter brings forth the contestations to this state of affairs by climate activists and grassroots mobilizations. By focusing on the tension between consensus and dissent, De Rosa analyses the strategies through which three activist groups repoliticize the climate in the context of Sweden's and Malmö's climate governance, detailing how they challenge the socio-ecological dominant order and how they contribute to the articulation of counter-hegemonic relations across scales.

Keywords: climate justice movement, direct action, grassroots mobilization, counter-hegemony

Malmö's Western Harbor in Sweden has been praised as one of Europe's most sustainable districts. Its status exemplifies Malmö's ambition to transcend the image of a derelict post-industrial city and become a symbol of the innovative green city. However, on closer inspection, the district appears to be an exclusive enclave of wealthy residents, while its alleged sustainability profile rests on the selective accounting of emissions and material and energy use, from construction to residents' lifestyles (Holgersen & Hult, 2020). Furthermore, according to a report on the local impacts of climate change (SKL, 2019), under the worst-case scenario of global warming at

Armiero, M., E. Turhan & S. Paolo De Rosa (eds), *Urban Movements and Climate Change: Loss, Damage and Radical Adaptation*. Taylor & Francis Group 2023
DOI: 10.5117/9789463726665_CH04

current rates of greenhouse gas emissions, Malmö's Western Harbor risks being submerged by the end of the century due to sea level rise and storm surges. As Holgersen and Malm (2015) put it, the "green fix" this district aimed to provide for the city is more of a "capital fix" divorced from its ecological and social basis.

The bitter irony of a brand new "sustainable" district at risk of being under water is more than a testament to the overtly narrow perspective that underpins most urban sustainability interventions. The district's likely underwater fate also serves as a reminder that the plethora of top-down climate actions currently being negotiated and implemented in Sweden and elsewhere may be irrelevant to climate stability as well as to climate adaptation—not to mention climate justice—if constrained within a reductionist, technocratic and market-oriented climate governance framework. Nonetheless, these interventions may still have an effect: they foster hopes that "something is being done" and give the impression of control by governments and of commitment by corporate actors, while channeling aspirations for climate stability into actions and policies that do not fundamentally challenge the warming trajectory, its structural drivers and the injustices it rests on, produces and exacerbates.

The perception of control by economic and political elites and their alleged commitment to addressing climate change are two expressions of what has been referred to as the *consensual regime of climate change governance*. Contrary to climate denialism, this broad and transversal consensus—spanning heads of states, corporate CEOs and international agencies and institutions—acknowledges the reality of climate change and mobilizes on the climate politically, yet articulating the climate as a political concern has the paradoxical effect of suspending the proper political dimension (Swyngedouw, 2010, 2011). Discursively, such a consensus portrays climate change as a universal human-made threat that makes "everybody" both victims and perpetrators, effectively silencing or displacing ideological and other constitutive social differences. Within the consensual regime, the solution to climate change is reduced to managing CO_2 in ways that are compatible with capitalist accumulation and growth, therefore leaving untouched the power of capital, the rule of the market and the global imperial hierarchies underpinning both. Examples of the post-political regime of climate change include the subsidizing of the fossil fuel industry by states ostensibly committed to mitigation, the framing of "energy transition" as electrification of the capitalist economy without reconsidering production and overall energy and material use, and relying on the profit motive to drive mitigation and adaptation responses, while omitting from the sphere of governance issues of accountability, fairness, compensation and the direct, nontokenistic involvement of affected communities. The overall effects include the rise of techno-managerial apparatuses driving questionable socio-ecological fixes (such as carbon markets, carbon offsets, "transition" fuels, etc.), the proliferation of new frontiers of extraction to feed transitions conceived as business opportunities and a growing reliance on future technological breakthroughs. The efforts mobilized in the name of the consensual regime of climate change governance thus not only fail to alter the warming trajectory and to provide due compensation to countries in the Global South for the climate debt contracted by developed countries, which would address rising losses and damages. These efforts also weaponize security concerns and "green" transitions to entrench further unequal relations, increasing risks for more vulnerable and less responsible countries and social groups (Perry, 2021), all in the name of "saving the planet." What appears as consensus thus conceals the hegemony of certain ways of framing, managing and addressing climate change that are instrumental to the perpetuation of the current socioeconomic status quo (Kakenmaster, 2019; Mouffe, 2000).

This state of affairs is increasingly being contested by various instances of climate activism and grassroots mobilizations exposing its insufficiency, depoliticizing strategies, false solutions and lack of concern for social inequality and vulnerability (Kenis, 2021; De Moor et al., 2021; Han & Ahn, 2020; De Rosa, 2022; Berglund & Schmidt, 2020). Climate activists disrupt the consensual governance of the climate by bringing back dissent and disagreement as the true engines of politics (Machin, 2013). Dissent expressed through political performances exposes and contests the process of depoliticization and provides platforms to marginalized voices and positions, inviting us to rethink the political in the era of climate breakdown.

In this chapter, by focusing on the tension between consensus and dissent, I analyze the strategies through which activists repoliticize the climate in the context of Sweden's and Malmö's climate governance. My investigation asks how climate and grassroots movements challenge the socio-ecological dominant order and to what degree their activism contributes to the articulation of counter-hegemonic socio-ecological relations across scales.

I concentrate on three activist groups that are partly overlapping and contiguous. What distinguishes them is the emphasis each places on a specific strategy, a relevant parameter for reflecting on various iterations of activism. The strategies in question are: (1) *policy pressure through civil disobedience*, used by Extinction Rebellion Skåne to challenge the national approach to mitigation; (2) *direct action and blockades against fossil fuels*, used by Fossilgasfällan to oppose the expansion of fossil gas in Sweden and the framing of it as a "transition fuel"; and (3) *building autonomy and self-organization*, the foundational principles of the Malmö-based social center, Kontrapunkt. This investigation offers insights into the workings of climate politics on the ground as well as providing useful lessons on how movements can influence climate policy and broader transformations. My main goal is to think *with* movements and theory to consider how bottom-up organizing can break the climate consensus deadlock and advance transformative climate action.

The research is based on six months of engagement with climate and social justice activists in Sweden from July to December 2019, on document analysis of grassroots campaigns and on semi-structured interviews with 13 core organizers (five identifying themselves as males and seven as females) of climate and social movements in Malmö, Gotebörg and Stockholm, with follow up interviews throughout 2020. The data were analyzed using manual coding and organized in thematic blocs reviewed in light of the main research questions. I also attended several events and direct actions as an active participant and as an observer, in line with my commitment to principles of equity and justice as a researcher and teacher, inside and outside academia (Green, 2020). This chapter is organized as follows: first, by drawing upon research in urban political ecology, social movements and climate justice, I elaborate my theoretical approach; second, I introduce Sweden's and Malmö's climate politics and highlight their critical features; and third, I focus on XR Skåne (XRS), Fossilgasfällan (FGF) and Kontrapunkt (KNT) to map their repoliticization strategies. Finally, I provide an analysis of my findings in order to contribute to current debates on the political relevance of grassroots climate activism.

Climate consensus, politicization and the role of social movements

Over the last 30 years, and increasingly since the Paris Agreement in 2015, the climate rhetoric of governments, economic actors and NGOs in institutional settings has nurtured the hopes of the global citizenry that top-down solutions were being devised and swiftly implemented. Against the obvious deceptions of climate denialism, the climate action they promised seemed like a beacon of hope. Nevertheless, despite the plethora of climate ambitions, goals and talks, this can be best characterized as "active inaction." Joint efforts to reduce emissions have not provided tangible results (Ciplet et al., 2015; Stoddard et al., 2021), while fair compensation mechanisms for losses and damages from climate-exacerbated disasters are still lacking. Yet, the burgeoning political and economic activity around the climate has been instrumental in capturing the debate around solutions to climate change and producing a consensual regime of climate change governance.

For Swyngedouw (2010, 2011), this regime is an outcome of the current post-political condition that forecloses politicization and neutralizes dissent through apparent participation and technocratic expertise in the context of an undisputed market-based socioeconomic organization. The result is that the climate becomes an object of policy, but only within the contours of the status quo. In practice, the kind of neoliberal environmental governance that emerges from the post-political order tends to reduce sustainability and climate issues to technical questions to be dealt with through techno-scientific and economic rationality, and through state-controlled procedures of so-called "participation."

This approach disavows and displaces social conflict in the name of climate change as a global humanitarian cause, something "to fix" without articulating specific political programs or socio-ecological projects, but only techno-managerial and behavioral transformations organized within a neoliberal capitalist order that lies beyond dispute. Reducing the entire issue of climate change to managing CO_2 is one of the main effects of the post-political approach. A fetishist disavowal that finds its completion in the making of CO_2 into a tradable and exchangeable commodity, allowing the phantasmagoria of the accounting practices of carbon markets, carbon offsets and abstract net zero ambitions (Carton, 2020; Watt, 2021).

In this context, consensus does not merely suggest agreement, but is an expression of hegemony. A consensus-based conception of the political and a singular conception of climate change—as manifested by the global regime of climate change governance—prevent the articulation of conflicting interests that do not fit dominant framings (Kakenmaster, 2019). The result is the reproduction of hegemonic narratives that thrive on the invisibilization of disagreement. Although it presents itself as an attempt at politicization, the consensual regime of climate change governance is in actuality a force of depoliticization. In other words, the environment is politically mobilized, yet this political concern suspends the proper political dimension. Thus, the question is no longer one of "bringing environmental issues into the domain of politics." Instead, it becomes "how to bring the political into the environment" (Swyngedouw,

2011, pp. 254–255). The political, in this perspective, is precisely the eruption on the climate and environmental stages of those knowledges, values and desires excluded by the consensual regime of climate governance. Going against consensus means to "open spaces of radical disagreement and rearticulation of other histories, other ways of being, and other 'solutions'" (Ernstson & Swyngedouw, 2018).

This is precisely what may emerge through the political performativity of subaltern experiences embodied by grassroots practices of politicization, resistance and self-organization, those challenging top-down programs of planetary management based on the narratives of "saving nature" and "saving the planet." A genuinely political recasting of the conundrum of global warming and the incipient socio-ecological collapse asks instead: Whose natures and whose planet? In Swyngedouw's words:

> [T]he political is the contested public terrain where different imaginings of possible socio-ecological orders compete over the symbolic and material institutionalization of these visions.... [It is] a terrain that makes visible and perceptible the heterogeneous views and desires that cut through the social body. (Swyngedouw, 2014, p. 90)

The reemergence of the political implies new practices of living together through the reappropriation of space (symbolic and material) and the production of new socio-ecological relations. Therefore, against a hegemony dressed as consensus, the political is about producing counter-hegemonic socio-ecological discourses and practices (Kurtz, 2020) with the potential to carve an expanded space for the imagination of alternative futures (Kenis and Mathijs, 2014). By reframing the climate debate in line with this conception of the political, my inquiry asks how climate and grassroots movements challenge the socio-ecological dominant order and to what degree their activism contributes to the articulation of counter-hegemonic socio-ecological relations across scales.

It is becoming clear that the required changes cannot be accommodated within a framework predicated on free-market fundamentalism and geared at profit and endless growth. Rather than a techno-managerial or a market adjustment problem, more radical positions—ranging from demands to "follow the science" by mainstream climate movements to "system change not climate change" by climate justice coalitions—regard climate change as an issue to tackle through the overall transformation of the socio-environmental assemblages and relations produced by the hierarchies and institutions of neoliberal capitalism.

Climate activism and politicization

Broadly united by demands for reductions in carbon emissions, climate movements are anything but a homogeneous phenomenon (Dietz & Garrelts, 2014). In recent years, the eruption onto the climate political scene of direct-action movements adopting nonviolent civil disobedience (like Extinction Rebellion and Ende Gelände) and of youth movements organizing mass demonstrations (like Fridays for Future), has made climate activism widely popular, channeling demands for reforms and wide-ranging transformations, especially in the Global North. In Sweden, both Fridays for Future and Extinction Rebellion, and their more recent offshoots, have been prominent in mobilizing people to demand more rapid and just climate action. In such a context of renewed citizens' advocacy, multiple strands of the climate movement have also advanced principles and criteria for accountability and fairness in climate change responses and pushed for inclusive and rights-based interventions (Krause, 2018; De Moor, 2021; Boda & Jerneck, 2019).

Grounded in more radical analyses and demands, climate *justice* emerged from the backdrop of decades of grassroots organizing to fight capitalism, environmental destruction and fossil fuels use (Temper et al., 2020; Gonzalez, 2020; Martinez-Alier, 2002), connecting environmental justice struggles by minorities and racialized groups with civil society mobilizations around the "climate debts" of the Global North toward the Global South (Bruno et al., 1999), and was formalized as a set of demands and principles by the counter-summits in Bali (2002) and Cochabamba (2010). Climate justice activism and critique has drawn attention to the historical and contemporary inequalities that underlie climate change (Schlosberg & Collins, 2014) and has politicized mitigation and adaptation responses that fail to address urgency, vulnerabilities and intersectional injustices (Walker, 2012), linking climate change to a wider critique of neoliberal globalization (Ajl, 2021), to imperial modes of living (Brand & Wissen, 2021) and to persistent colonial structures (Red Nation, 2021). The manifold antagonistic framings employed by climate activists counteract attempts to construct climate change as a "post-political" issue and support platforms for anti-systemic approaches against the elite capture of the climate crisis (Chatterton et al., 2013; Rice et al., 2021). This approach has influenced pressing debates around loss and damage, mitigation paths and reparations (Boyd et al., 2021; Abimbola et al., 2021).

Climate activism has been portrayed as being focused mostly on mitigation (e.g., Dietz & Garrelts, 2014; De Rosa et al., 2022) and engaged in a global perspective that is disconnected from local justice issues. However, environmental and climate justice movements have helped to situate concerns for climate justice in local socio-environmental struggles (Sikor & Newell, 2014). Social justice activists also play vital roles in connecting various urban struggles by enabling alliances between more and less resourceful groups (Hansen, 2020), thus contributing to new spaces of convergence where vulnerable communities and climate activists join efforts to address local adaptation-related injustices and promote transformative approaches to urban adaptation (De Rosa et al., 2022). This happened in New York City after Hurricane Sandy, when the organizational capacity of Occupy Wall Street was reoriented toward disaster-relief, giving rise to Occupy Sandy (Dawson, 2017). Similar dynamics occurred in post-Katrina New Orleans (Bullard & Wright, 2009; Crow 2014). While disaster response is not the same as adaptation, more recent studies document adaptation planning emerging within the organizational and conceptual horizon of climate, environmental and social justice movements (e.g., Méndez, 2020; Climate Action Lab, 2019).

The contribution of social movement-led politicization of the climate has been the crucial staging of the demands, experiences and desires banned from the consensual regime of climate change governance. Moreover, such instances of politicization are providing physical and discursive platforms of convergence for broader transformative political projects (Newell et al., 2021).

I now turn to how the dynamics of depoliticization and repoliticization are playing out in Sweden. In the following section, I provide an overview of Sweden and Malmö in the context of the climate crisis, before delving into the findings and analysis of the three movements' repertoires of action.

Sweden and Malmö in the climate crisis: Policies and impacts

Sweden is often hailed as a country leader in climate action. According to the Climate Change Performance Index 2021, it scores better than 57 other countries and the EU in climate protection efforts and progress (Burck et al., 2021). The country also aims to become the world's first fossil-free welfare state by 2045.[1] However, Sweden is also routinely criticized for the insufficiency of its climate action and for embracing controversial climate solutions by a rising generation of activists that, through strikes, campaigns

1 As stated in the Sustainable Development Goals Partnership Platform of the United Nations, https://sustainabledevelopment.un.org/partnership/?p=33918.

and direct action, has helped to mainstream awareness of climate breakdown and knowledge about its scientific basis and unfolding effects.

Sweden aims to reach net-zero emissions of greenhouse gases (GHGs) by 2045, followed by negative emissions. This target covers all emissions within Sweden's borders, the so-called territorial emissions, and does not include emissions from consumption, from international transport and from land use, land-use change and forestry (LULUCF). The emissions of hard-to-abate sectors (like agriculture) will be offset by so-called supplementary measures, such as increased carbon sinks, bioenergy with carbon capture and storage, or investments in climate change mitigation projects in other countries. After 2045, the supplementary measures should exceed the remaining emissions in order to create negative emissions. The goals and monitoring system are regulated by the Climate Policy Framework of 2017 and by the Climate Act of 2018. These policies also established the Climate Policy Council, an independent, interdisciplinary expert body tasked with evaluating the alignment of policy and action. Finally, in June 2020, the government established the Ministerial Working Group on Climate Policy to coordinate the implementation of climate action and the post-pandemic recovery. The nation's emissions were reduced by 2.5% in 2019, better than previous years when this reduction was about 1%, but still far from the 6 to 10% per year needed to reach net-zero emissions by 2045, and nowhere near the 10 to 12% needed to comply with the pace and with the equity principle of the Paris goals (Anderson et al., 2020). The latest Swedish Climate Policy Council's evaluation of the government's post-pandemic rescue packages finds that "the pace of climate transition remains too slow, and current policy is insufficient for achieving the climate goals" (SCPC, 2021, p. 13). Indeed, "only one-tenth of the Government's recovery efforts also contribute to achieving Sweden's climate policy goals" (SCPC, 2021, p. 13).

Too slow an emissions reduction rate is not the only issue. Sweden has high levels of resource use and CO_2 emissions in consumption-based terms. A sustainable level of resource use is about 7 tons of material stuff per person per year. Swedish people consume on average 32 tons per year.[2] This points to a lack of consideration for issues of scale. The Swedish consumption-based GHG emissions are 82 million tons, and about 57% of these emissions occur in other countries as a result of Swedish consumption. The climate-affecting emissions that arise because of Swedish imports constitute a large and

2 According to the Sustainable Development Index, https://www.sustainabledevelopmentindex.org/.

growing share of Sweden's total consumption-based emissions (Almqvist et al., 2020).

When we turn to impacts, Sweden will become warmer and wetter in the coming years (SCCV, 2007). The average temperature is expected to rise three to five degrees more than the global average by the 2080s. Precipitation will increase in autumn, winter, and spring, while the summer will be warmer and drier (SCCV, 2007). Consequently, extreme heat and flooding will increasingly affect Sweden. Other climate risks include landslides, erosion, storm damage, drinking water contamination, heat waves, drought, spread of diseases and challenges for reindeer herding.

A sharp increase in temperature extremes in Sweden because of climate change has been already observed. Research on the Stockholm area has linked the substantial warming of recent decades to an increase in mortality rates (Åström et al., 2013). The region of Skåne in the southernmost part of Sweden, where Malmö is located, is forecasted to be one of the hardest hit by the effects of climate change (Hall et al., 2015). It has the largest share of buildings, infrastructure and beaches at risk from flooding, coastal erosion and sea level rise (SCCV, 2007), and it is among the country's regions facing the greatest increase in risk of forest fires (Yang et al., 2015). In recent years, these impacts have become increasingly noticeable. Drought, heat waves and forest fires were experienced during the summer of 2018, leading to more than 600 excess deaths across Sweden (Åström et al., 2019). Likewise, a series of floods and storms have had major impacts on Sweden during the last decade, such as the flooding event of 2014 in Malmö, a city classified as being at considerable flood risk in the near future (Haghighatafshar et al., 2014).

In the absence of national-level adaptation governance, adaptation largely falls to municipalities. However, research has shown that municipalities' ability to deliver on this front is limited by a lack of support from national and regional authorities, and by a lack of collaboration (Dymén and Langlais, 2013). Malmö's adaptation plans furthermore lack considerations of social justice issues (Nylund, 2014; Brink & Wamsler, 2018; De Rosa et al., 2022). Brink and Wamsler's analysis (2018) of citizen–municipality interactions for adaptation in three Swedish municipalities, including Malmö, found that interactions often exclude or disadvantage vulnerable groups and fail to address equity in adaptation.

To sum up, Sweden—and Skåne, in particular—is already experiencing considerable climate impacts and lacks a coordinated and justice-oriented approach to mitigation and adaptation. Even though the country is considered a climate leader, it is not fulfilling its Paris Agreement obligations

(Anderson et al., 2020) and is thereby contributing to locking in further climate disruptions.

Grassroots climate activism in Malmö and Sweden

Malmö has a long history of social movements and grassroots political organizing. In recent years, climate activists have mobilized in and through this city to build local coalitions and to channel local energies toward (inter) national networks and actions. I now turn to an exploration of the ways in which Swedish activists politicize the current governance of climate change, their claims and strategies, and their relations with transformation.

XR Skåne

Extinction Rebellion Skåne (XR Skåne) focuses on counteracting the insufficiency and "false solutions" of national and municipal climate policies by employing a main strategy of political and social pressure through persistent acts of nonviolent civil disobedience. About 70 activists from XR Skåne inaugurated the birth of the local XR node on December 14, 2018, by occupying a street in Malmö city center and halting traffic. For most of the participants, this was their first experience of activism. Disrupting traffic flows in cities is one of XR's main strategies for drawing attention to their three demands to governments: tell the truth about the climate crisis, achieve zero emissions by 2025 and create citizens' assemblies to drive the transition. This first action in Malmö took off thanks to a core group of activists trained by previous experiences of climate and environmental organizing who replicated the XR UK blueprint.

Between December 2018 and June 2021, XR Skåne performed about 60 actions in Malmö and beyond, including roadblocks, disruptions of municipal council meetings, die-ins at shopping malls, artistic interventions, airport occupations, demonstrations and cycling swarms. They also provided training and weekly information meetings for new cadres. As part of networks and coalitions, activists from XR Skåne organized local actions in conjunction with the weeks of uprising coordinated by XR globally in 2018, 2019 and 2020, and joined blockades against fossil fuel infrastructure and forest destruction organized by climate justice movements in Sweden, Denmark, Norway and Germany.

In March 2019, XR Skåne launched one of its main initiatives: getting Malmö municipality to declare a climate emergency. According to the activists, declaring a climate emergency at the local level would speed up the transition, prepare for unavoidable shocks and provide a means to influence higher levels of government, rather than merely being an objective in itself (Interview with J.). It would put the basis for prioritizing a climate emergency plan in the municipality's strategic work to build up competences, to take compensatory responsibility for a quick transition until senior politicians take full responsibility and to inform all Malmö residents about the climate emergency. XR activists used the "Malmö Initiative," the city's own platform for allowing citizens to submit proposals that must be discussed by the municipal council if at least 100 people vote for them. XR Skåne's climate emergency proposal was signed by more than 800 people, making it one of the initiatives that received the most votes since the system was introduced. The issue was thus debated by the municipal council on January 13, 2020, but instead of *declaring* a climate emergency the council only *acknowledged* that there is one. According to one politician, the risk to commit as a municipality to take compensatory responsibility for the failures at a higher level was considered "financially impossible" (Rosén, 2020).

In April 2019, XR Skåne collaborated with other branches of XR in Sweden to produce and publish a pamphlet in English and Swedish with the title, *When the Climate Crisis Comes*, addressed to Swedish politicians. It was an answer to a brochure distributed in 2018 in English and Swedish by the Swedish Civil Contingencies Agency advising the population of Sweden on how "to become better prepared for everything from serious accidents, extreme weather and IT attacks, to military conflicts" (SCCA, 2018, p. 3). Even though extreme weather events due to climate change were mentioned, all the recommendations referred to the threats of war and terrorist attacks. XR therefore decided to answer by making a guide with the same design, but with a focus on the climate crisis (XR Sweden, 2019). It illustrated the risks of failing to confront the human-induced warming of the atmosphere and ecosystem collapse. On page 4, the 2018 heat wave and its effects on Sweden (persistent drought, increased wildfires, reduced harvests and excess deaths in cities due to heat) are explicitly mentioned in order to show "the consequences of not acting now." The evocation of extreme weather and ecosystem collapse based on scientific projections and emotional messaging is a central feature of XR's framing. By grounding the global climate and ecological crises in the national context with references to recent extreme weather events in Sweden, XR brought the climate crisis nearer to the experience of ordinary people. However, the mention of the heat wave of 2018, as well as the flooding of Malmö in 2014, served the function of reinforcing the overall XR message rather than innovating the focus of

XR climate activism substantially (e.g., through calls for just adaptation or compensation). Another notable element in the XR pamphlet is its call to Sweden to take the responsibility for its *total* climate impact, i.e., the "historical responsibility for already accumulated emissions, and ... ongoing high emissions per capita" (XR Sweden, 2019, p. 6). Such an awareness of global imbalances in responsibility and vulnerability calls into question the exclusive focus on national climate policies on territorial emissions and demands forms of accounting and action in line with the emissions linked to past and present consumption of imported goods, international transport and from Swedish financial and economic actors abroad. The focus on consumption was also at the center of several die-ins in shopping malls and department stores, such as the one on November 29, 2019 at the Emporia department store on Black Friday, which drew attention to the oversized consumption footprint of Swedish people.

Since its first action, XR Skåne has continuously organized disruptive interventions in the urban fabric, using persistence as one of its signature tactics. It has joined Fridays for Future in its weekly strike in front of Malmö municipality building and has provided support to other campaigns and movements such as Fossilgasfällan's action in Göteborg.

XR Skåne aims to continuously grow its activist base through training and information meetings. However, so far it has been joined almost exclusively by white middle-class citizens. According to J., the meetings it organizes "make knowledge about the crisis and the solutions a mobilizing factor" (Interview with J.). A great deal of attention is devoted to communication by filming actions, spreading training videos and holding online and in-person meetings to socialize and cultivate relationships beyond activism. Finally, XR Skåne is leading a national campaign, but also experimenting with more local issues and stimulating a change of perspective among participants. As M. stated:

> One of the framings when we organize training is that we build affinity groups to prepare for a really bad future, and also for fascist regimes. Many people are not ready to hear that, but I use that framing in the beginning when we explain why we need affinity groups, not just for the action, but because we need to get organized. (Interview with M., XR Skåne)

Fossilgasfällan

Fossilgasfällan (The Fossil Gas Trap, FGF), a climate justice movement explicitly oriented toward anti-imperialism and anti-capitalism, utilizes more confrontational direct actions to blockade fossil fuel infrastructure and establish climate camps, cooperating with national and international coalitions. A group of activists based in Göteborg with backgrounds in animal rights, migrant justice, Sámi rights and environmental NGOs created FGF in 2017. It started as a grassroots campaign working for a "world where fossil fuels belong to history and where the burden of dealing with the consequences of the climate crisis is fairly distributed." Its members focused on fighting the normalization of and consensus around fossil gas as a "transition fuel" for climate neutrality in Sweden and Europe, targeting the project of a gas terminal expansion in the port of Göteborg by the company Swedegas. Throughout its three-year campaign, it received support from the Gastivists, a global network connecting frontline communities with climate justice groups to oppose new gas infrastructure, facilitating knowledge creation and providing a practice-based model of intervention rooted in civil disobedience through workshops and training.

The main point of contention of FGF was the framing in policy circles and in the energy sector of gas as a bridge fuel for the energy transition required by climate change. Starting by dismissing the term "natural gas" for "fossil gas," their analysis showed how the rising popularity of gas was the result of a greenwashing operation by fossil fuel companies and their lobbies in government. The main tools for building the campaign were social media, articles in the press and actions in cities. Through a dedicated website, FGF set up a "knowledge bank" collecting sources challenging fossil gas and offering instruction about alternatives. Moreover, virtual info meetings provided an easy way to broadcast information, to gather support and to expose the financial interests and players behind the Swedegas project.

Informed by climate justice concerns, FGF took seriously the claims and experiences of local communities impacted by the extractive processes used to exploit fossil gas, most notably through hydraulic fracturing, also known as "fracking." This led to a holistic critique of fossil gas associated with the development of a set of alternative proposals for the transformation of energy systems and of society. In one of their first public actions, on February 11, 2018, about 30 people gathered outside the Social Democrats' office in Göteborg to protest the construction of the terminal. In a fiery speech, the activists set the terms of the issue within the broader geographies and consequences of extraction:

> Fossil gas has major negative impacts where the gas is extracted and the infrastructure is built: water poisoning, earthquakes, landslides, repression of human rights, destruction of land areas and agricultural

> areas. What is happening now in Göteborg harbor affects all of Sweden and all of the world's people. (Speech by K.)

The campaign by FGF gave the climate movement in Sweden a clear target around which a broad alliance of activists and concerned people could coalesce. To ground broad, complex and (seemingly) abstract issues like climate change by targeting specific nodes and facilities implicated in the fossil economy, where opportunities for short-term gains and victories sustain momentum and movement building, has been an explicit strategy of the climate justice movement for at least a decade (Cheon & Urpelainen, 2018).

By forging ideal and practical links with frontline communities resisting fossil fuel operations and expansion, FGF was able to frame its actions as both an ecological and social battle with immediate implications for the lives of communities far away from Sweden and as a decisive step toward halting GHG emissions. This internationalist approach was operationalized by inviting representatives of communities impacted by gas extraction and infrastructure from the USA, Argentina and Ireland to Sweden. Their counternarratives on fossil gas merged the claims of place-based campaigns for environmental justice, indigenous sovereignty and rights to a clean environment with the climate struggle against fossil fuels expansion in Sweden.

FGF aimed to put the task of debunking fossil gas as a climate-friendly transition fuel on the agenda of the climate movement and of Swedish climate debates. The central objective of their organizing was to halt the completion of the terminal and to strengthen the climate justice movement in Sweden and Europe by "learning from each other [and by] building on and making use of the momentum we collectively created" (Interview with A.). After forms of dialogue to draw media and political attention were exhausted, the coordinated direct action to block the terminal became inevitable. Together with the growing network of European direct-action movements against fossil fuels and enlisting social justice movements from all of Sweden, FGF cocreated the campaign Folk mot Fossilgas (People against Fossil Gas), at the same time a coalition, an event and a climate camp geared toward blocking the gas terminal's operations for at least one day on the weekend of September 6–8, 2019.

Folk mot Fossilgas was launched six months before the event through an international call inviting "anybody who wants to act for climate justice and who agrees with the action consensus" to join the Swedish activists at the port of Göteborg. Action consensus is a typical mode of involvement utilized by direct-action movements against fossil fuels in Europe to ensure informed participation by anybody who is sympathetic with the movement's goals and tactics, and to allow the widest participation irrespective of ideological adherence to a specific movement's identity (Vandepitte et al., 2019). The kind of consensus sought by climate and environmental justice activists does not silence antagonism and partisanship. Instead, it exposes them by drawing a line against racial, gender and other forms of discrimination, and by grounding collective action in anti-fascist and anti-capitalist values.

As the date approached, the European network of allies organized meetings and training sessions in Malmö, Copenhagen, Berlin, Amsterdam, Brussels and other Northern European cities. The meetings recruited participants, while the training introduced the neophytes to civil disobedience and to the principles of coordinated direct action. The training was a crucial element for the success of the action, transferring practical skills and building a sense of empowerment and safety among participants, preparing them to respond to police intervention through nonviolent civil disobedience and providing coordinating skills to manage decisions within affinity groups to dismantle the groups' internal power dynamics.

On September 6, 2019, people began converging on the camp set up by the organizers in a meadow about a kilometer from Göteborg's port. The camp included a big circus tent for the assemblies and five smaller structures for the info point, a dining area, a media center, legal support and medical support. The media center worked tirelessly to film, edit and upload content, to produce and disseminate articles and press releases, and to monitor the mainstream media narrative of the blockade before, during and after the event. Interviews with international delegates on video and in crowded plenary sessions brought to Göteborg the concerns and the solidarity of frontline communities fighting against gas extraction and gas terminals around the world.

The purpose of the media center was to make sure that the narrative of the campaign reached the broader public. Since mainstream media and politicians often dismiss civil disobedience, the "narrative battle" was a key site of resistance. The activists waged it effectively, as shown by the fact that spokespersons from the campaign were interviewed by most of the Swedish major media. The practical tasks of sustaining the camp's communal life were distributed on a voluntary basis and taken seriously by volunteers. The participants' average age was under 30, and the overwhelming majority were European and white. Time passed between assemblies, food sharing and training. The training at the camp, organized by the Climate Justice Program, focused on tactics of civil disobedience and coordinated mass action, the distinctive tools of the militant climate movements consolidating in Europe. Essentially, these kinds of actions are mass interventions on the operation of physical infrastructure implicated in the fossil economy with the aim of slowing it down or blocking it altogether. The objective is not sabotage, but a demonstration of grassroots force that aims at maximum visibility to nurture imaginaries of revolt and to inspire emulation. As acts that defy the law by invading private property or occupying public places, the risk of being arrested is a foregone conclusion. However, larger numbers of participants tend to guarantee collective security. When it goes well, the result is the collective construction of a symbolic event, highly visible and easily understandable to the outside, inviting and safe for inexperienced activists, culminating in a formative and transformative experience for individual participants.

On the morning of September 7, about 500 people blockaded the four gates of the terminal by sitting close together on the pavement. The police officers stood still. On both sides of the gate, the files of tanker trucks were blocked. In the meantime, drones were buzzing around filming video for the action organizers. Two people in yellow vests with the role of communicating with the police spread the news that they were not going to remove the activists. The tension disappeared, and the same news arrived from the other blockades. Several television channels carried out interviews and talked about "fossil gas" instead of "natural gas," a clear symbolic victory. In the end, the blockade lasted about 12 hours, achieving the temporary interruption of the terminal's output flow, positive media coverage and no violence by the police. About a month after the action, the Swedish government announced that it had denied Swedegas the final permit for expanding the terminal. Moreover, expansions and construction plans for gas terminals in several Swedish ports were halted. While it would be a mistake to overemphasize this single action, it contributed to fracturing in multiple ways the widely held commonsense view of Sweden as a climate leader, representing a turning point for the climate justice movement in Sweden and neighboring countries.

Kontrapunkt

Kontrapunkt, a social center based in Malmö, focuses on struggles for justice and democracy, and on cultural work. Its main strategy is based on building self-organization and autonomy. A local group of activists started it in 2006 as a social movement rooted in the Malmö neighborhood of Möllevång "using culture as a tool for organizing a community" (Interview with L.).

The founders took anti-discrimination, mutual aid and social justice as their guiding principles. In 2010, inspired by the social center movement in Europe and by community movements in North and South America, the activists rented a large warehouse and turned it into a social center: Kontrapunkt. With the stated aim of deepening democracy by taking power back to the people, they expanded their initial aims to engage in several forms of service provision and political organizing. Through their initiatives, they have stirred anti-gentrification campaigns and solidarity-based work with homeless people, refugees and Roma migrants (Hansen, 2020).

In one instance of grassroots activism and self-organization, their contribution was crucial in providing relief and support to thousands of refugees, mostly from Syria, who arrived in Malmö in 2015. During that year, about 100,000 people passed through the city, but national and local authorities lacked a commensurate organizational response. Civil society organizations stepped in to provide immediate help, and Kontrapunkt was at the forefront of this effort. For 135 days, more than 17,000 people were provided with beds, food, clothing, health care, transportation and counseling, each of these activities supported through a variety of fundraising initiatives.

These events happened in a moment of self-reflection for Kontrapunkt and triggered a crucial realization:

> In August 2015, we left for a weekend in the woods to discuss internal political matters. We had 3,000 square meters in the city, a bus, a circus tent, all kinds of different kitchen materials, and we were struggling to maintain everything. So, we asked ourselves, "What are we going to do with all these things?" And we answered, "We are an infrastructure. Things will happen in this world, and we're gonna be ready." Then we came back on Monday, and refugees were walking to us. It was literally crazy. Now I feel and understand what we have been doing, what we have been fighting for. This is why we exist. So definitely this is part of our mission or vision of what we want to do or be in the future. (Interview with M., KNT)

The attitude of the municipality of Malmö toward the self-organized reception of refugees by Kontrapunkt was largely hostile. Formal collaboration was never established and funding was not provided, even though the refugees seeking shelter were informally directed by municipal officials to Kontrapunkt. When the municipality set up its own shelter, Kontrapunkt was threatened with a fine of a half a million Swedish crowns (about €46,000) for lacking the permits to have people sleeping on its premises, and it had to stop providing this service. Nevertheless, this experience reinforced the commitment to enhance Kontrapunkt as a political and social space aimed toward justice and autonomy:

> This is often how we describe ourselves, as an infrastructure for groups, for capacity building, a resource center for people who want to create something or mobilize. We have a lot of things in here, both competence, but also workshops, so you can build things, rehearse, have a place for meetings or for performances. (Interview with M., KNT)

Despite the wide spectrum of activities Kontrapunkt engages in, climate concerns appear, at least on the surface, to intersect with its work only tangentially. The main concerns of the variegated community that animates Kontrapunkt—made up of people with many different backgrounds, in some cases previous refugees who were supported by the social center and then became themselves activists—revolve around migrant justice, anti-fascism, solidarity with the poor and disenfranchised, but also with other movements in the city. Kontrapunkt has welcomed XR and FFF activists, allowing them to use their facilities and supporting their actions on the spot with food, tents and other logistical needs. However, climate justice claims have not been put forward by Kontrapunkt's activists. Yet, rather than a conscious omission, this can be seen as a result of their focus on practical, social justice-oriented organizing. It is precisely this dimension of practice-oriented and justice-based activism that justifies a characterization of Kontrapunkt as an experiment that can be tied to strategies of preparation, disaster response and broader adaptation to climate breakdown in ways that could foster transformative pathways. In this sense, Kontrapunkt could provide the meeting point where vulnerable groups and climate activists can politicize local climate justice issues. However, given issues of scale, the work of Kontrapunkt can best be seen as part of a puzzle that also requires policy pressure and direct action to redress climate injustices.

Discussion: Grassroots strategies in the era of climate breakdown

The three movements I have introduced here disrupt in different ways the consensus around the climate governance of local and national authorities in Sweden through the politicization and contestation of the socio-ecological dominant order it upholds. In this section, I first distill the lessons from each movement separately for analytical purposes and then draw more general conclusions about the relationships between politicization and climate activism.

XR Skåne focuses on policy pressure through civil disobedience. It breaks the consensus around institutional responses to the crisis by highlighting the urgency required to avert catastrophic outcomes. Raising the alarm is its main concern. Its narrative relies on scientists' recommendations and attempting to substitute the government's consensus with the scientific consensus. One of XR's strengths is its ability to mobilize people with no previous political experience or explicit political leanings, including individuals who are worried about the climate, but do not know what to do. This appeal is designed by creating a nonsectarian environment and through an approach to civil disobedience that seems to based on solid research. The request for a municipal climate emergency declaration is a way to unsettle the Swedish self-perception of being a forerunner in climate mitigation and adaptation. The use of the "Malmö initiative" also signals XR's willingness to take advantage of all the available tools, legal and illegal (such as blocking roadways). Similarly, by hijacking the government's language and format for a pamphlet on disaster preparedness, XR Skåne engages in the work of counter-framing and mainstreaming climate concerns by grounding the planetary issue of climate change in local realities. Finally, by highlighting the entirety of Swedish people's emissions and by denouncing their consumption habits, as in the pamphlet and in direct actions at malls, XR debunks the reductionist accounting of climate emissions that does not square with reality. Yet XR lacks a vision of system change and consciously avoids the question of which kind of socioeconomic system should replace fossil capitalism, preferring to leave these issues to the deliberations of citizens' assemblies, which is one of its core positions.

The second group, Fossilgasfällan (FGF), aims to stop the expansion and use of fossil fuels by employing direct action and blockades. It attempts to change the state's narrative on gas as a "bridge fuel" by exposing the contradiction between expanding gas infrastructure and national climate goals. Its focus is on capitalism and neocolonial relationships as the main engines of social injustices and environmental destruction: activism exposes the capture of states by financial players, lobbyists and fossil fuel companies. By putting the spotlight on the entanglement of financial interests and the socio-environmental consequences of fossil gas along all stages, FGF highlights the struggles for land, water and other basic needs waged by communities impacted by gas extraction. Putting the focus on the material infrastructures that reproduce the relationship between the fossil economy and climate chaos disrupts the consensual framing of climate action as the management of disembodied emissions, thus welding emissions to a larger set of concerns. Like XR, the process of engagement in FGF happens through direct action, the safety of activism and a nurturing environment. This also breaks the narrative that individuals can only participate in climate solutions through consumption habits. By (re) creating spaces and communities around direct actions, they turn climate justice concerns into concrete practices of prefiguration. In particular, the life of the climate camp calls into question the pursuit of self-interest and competition (the hallmarks of neoliberalism) in favor of practices of mutualism and cooperation, making the experience of participation potentially transformative. The group's challenge to the dominant order also supports other environmental struggles in Sweden and elsewhere, and its explicitly anti-racist, anti-capitalist and anti-colonial agenda allies them with indigenous people. However, it could be argued that its approach is more symbolic than disruptive: the blockade and the occupation are not meant as long-term disruptions or sabotage of fossil fuel operations, but rather as demonstrations of the popular will through the force of grassroots organizing.

The third group, Kontrapunkt (KNT), is rooted in Malmö and engaged in a variety of local social justice issues. It aims to build autonomy and self-organization, challenging the representation of Swedish democracy as well functioning. Therefore, it works to build people's power from below in the day to day. Support for the needs of the most disenfranchised is front and center, not as charity, but in the name of solidarity. It has, therefore, a thorough knowledge of the ills and vulnerabilities of the local communities subjected to poverty, racism and social exclusion, and it maintains a large network with other activists. Adding this to its material base and logistical capacities, KNT can act as a social infrastructure for mutual aid that could serve both as a means of disaster response and the basis for transformative adaptation. Becoming a space of convergence for different grassroots political groups is a task that urban social centers may be uniquely positioned to perform, but only by explicitly integrating issues of climate and social justice into their agendas.

The strategies and tactics presented above are united by the attempt at disrupting the consensual regime policing the framings of climate change governance and democracy in Malmö and Sweden. I provide an overview of their respective strategies, tactics and shortcomings in Table 4.1.

Table 4.1. An overview of strategies, tactics and shortcomings of the movements

	Key strategy	*Key tactics*	*Key shortcoming*
XR Skåne	Policy pressure through civil disobedience to achieve immediate emissions reduction, climate emergency declaration and citizens' assemblies	– Highlight urgency – Rely on scientific consensus – Nonsectarian involvement and training – Ground climate breakdown in the local context – Indicate Sweden's responsibilities	Lack of a vision of systemic change
Fossilgäsfallan	Stop the expansion and use of fossil fuels through direct action and blockade based on anti-racist, anti-capitalist and anti-colonial politics	– Debunk narratives about gas as a "bridge fuel" – Expose states' capture by financiers, lobbyists and corporations – Put the spotlight on the socio-environmental consequences of fossil gas at all stages of its extraction and use – Target the actors and material nodes of the fossil fuel economy – (Re)create spaces and communities around direct actions – Support environmental struggles in Sweden and elsewhere	Direct actions may be more symbolic than concretely disruptive
Kontrapunkt	Build autonomy and self-organization to increase people's power from below	– Support the needs of the most disenfranchised in the name of solidarity – Maintain networks with other activist groups – Logistical capacities and a physical space for organizing – Potential to act as social infrastructure for mutual aid and disaster response	Climate concerns intersect only tangentially with its work

In the current context of worsening climate impacts and increasing vulnerabilities, the consensual regime of climate change is unwilling to put into question the market-based socioeconomic organization of societies dependent on capitalist growth, accumulation and imperialist hierarchies. Such a consensus is thus more correctly framed as a manifestation of the hegemony of neoliberal capitalism and as a force of depoliticization (Swyngedouw, 2010, 2011) that obliterates political conflict by covering the exclusion of alternative voices and approaches in techno-managerial language and the illusion of participation. New climate movements and sustained activists' groups drawing from traditions of resistance show the potential to interrupt the post-political order embodied by the consensual regime and to bring again to the center the political, conceived as the irreducibly conflictual staging of demands on the "part of those that have no part" (Rancière, 2010).

The task of repoliticization goes hand in hand with the opening of material and symbolic spaces for the articulation of dissent against the prevailing order (Ernstson & Swyngedouw, 2018; Kakenmaster, 2019). Such openings are essential to putting front and center the voices, experiences and desires of the social groups that are excluded by a hegemony dressed as consensus, in a gesture that inaugurates the political. In this sense, a movement like Extinction Rebellion tends to perpetuate some aspects of the post-political order precisely through its claim of being "beyond politics." Local nodes, as in the case of XR Skåne, seem willing to repurpose and to adapt XR's toolkit and frames, but hardly overcome the embracing of scientific discourse on climate change as categorical truths. This approach may reproduce depoliticization by deploying scientism. Moreover, even if shaking up some of the depoliticizing assumptions and self-perceptions regarding a country's climate performance (like emissions accounting in Sweden) provides some form of repoliticization, and even if the emphasis on urgency disrupts the trust in arbitrary timelines of required emissions reductions, an activism that relies heavily on science and pretends to act for the whole of humanity risks seeing its message co-opted and distorted. By creating the conditions and the demand for muscular state interventionism in climate issues, mobilizations that lack a radical critique of the unsustainability and injustice embedded in the hegemonic system of neoliberal capitalism may find their demands integrated in what Blühdorn and Deflorian (2019) define as the "collaborative management of sustained unsustainability."

The activist work of exposing the connections and relations across different scales that are systematically hidden in the consensual regime of climate change governance remains an effective repoliticization strategy (Derman, 2020). By making visible the links between spaces of extraction, consumption and climate impacts, activists debunk reductionist accounts of the causes and consequences of climate breakdown and the narratives of "solutions" foregrounded by the elites. It is a process of complexifying the terms used to interpret climate change, providing a narrative of historical depth, spatial extension and social implications vis-à-vis an approach solely focused on CO_2 accounting and substitution. While XR lacks the emphasis on justice and awareness of the role of capitalism in multiple crises, or avoids talking about it, FGF puts these front and center, while KNT faces them in its daily political work.

Another noteworthy aspect of current instances of climate organizing, even though specific to Western, capitalist-core countries, is the politicization of personal action, shifting it from individual consumer choice (a post-political identity) to personal engagement to achieve systemic change. The willingness to self-sacrifice exhibited by the activists using direct action, ready to be arrested or to break the law in the name of fixing climate and achieving justice, also by leveraging the relative privilege of mostly having white, middle-class and European backgrounds, exemplifies an idea of climate engagement radically different from the consensual image of the "green lifestyle" as the accepted manifestation of climate consciousness. Instead, a more combative and confrontational stance against a system unwilling to change is depicted as brave, honorable and even necessary. Moreover, such self-sacrifice and the consequences it engenders highlight the function of training, political formation and individual empowerment climate movements promote. Against the hopelessness of witnessing the crumbling of societies and ecologies, training prepares and expands the ranks of the emerging cadre of activists and provides them with conceptual and practical tools for resistance. Given the prospect of rising eco-fascism and impending disasters, a broad and trained generation of activists ready to act and to radicalize climate action may be an antidote to both hopelessness and negative developments. This element is also connected to the potential of constructing bottom-up social infrastructure for adaptation and disaster response and makes us reflect on the limits of current legal systems to accommodate those actions—disruptions, blockades, free aid to the vulnerable—that are much more in line with the severity and danger of the climate crisis. On the other hand, a grassroots strategy based on disruptive actions to impose demands on governments that is not accompanied by a parallel work of construction and diffusion of appealing political, economic and social alternatives for the citizenry may engender polarization within societies, which in turn could be leveraged by the elites to increase the repression of climate activists.

In our current conjuncture, climate activism may be best positioned for creating political platforms of convergence between multiple issues. Politicization triggered by climate concerns necessarily traverses fields of social action and existence as diverse as labor relations, human rights and Global North/Global South unequal dependencies. Yet this can happen only by going beyond the fixation with carbon and emissions tied to the agenda of the global regime of climate change governance that some movements display. That agenda, and the consensus around it, need to be overcome rather than embraced. The task of politicization is still to bring issues from the margins to the center without compromising the radicality of performative political demands. More importantly, climate activists' work is also about building spaces of counter-power and counter-hegemony, something that can only be achieved by threading the climate struggle together with other environmental, social and political movements.

Conclusion

A crucial synthesis emerging from my analysis is that social justice movements and climate movements need each other to politicize climate injustices and to contest the socio-ecological dominant order. While social centers and solidarity-based activism have the capacity to respond to emerging needs in vulnerable communities, and thereby to identify local shortcomings as well as potential solutions, the growing climate movement is well positioned to build policy-oriented pressure on local and national governments to assume their responsibilities and implement solutions at a more encompassing scale. Such alliances will be decisive for developing the social pressure and organizational capacity needed to drive broader socio-ecological transformations.

While I am writing these words, the movements analyzed here and other emerging groups continue to experiment with tactics, strategies and claims, exploring new directions of convergence, coalition-building, political pressure and self-organization. Promising venues of research could deepen the understanding of the manifestation of political in climate mobilizations by focusing on the radical discontinuities they engender in the "order of the sensible" policed by the elites' hegemonic approach to global climate governance. In particular, the recent focus of climate movements in the Global North on "canceling the debt" of countries in the Global South speaks precisely to the politicization of the historical legacy of colonialism and extractive capitalism excluded by mainstream climate discussions. Another promising area of research is the increasing academic activism of universities' research and teaching staff, joining and organizing civil disobedience actions against banks, financial actors, government ministries and their own academic institutions, advancing claims of varied radicality. Finally, cutting across climate mobilizations of different kinds and scopes, there is an urgent need for more research on tactics and strategies that can change the balance of forces quickly enough and support the building of counter-hegemonic power from the grassroots. That is, the *how* of transformation is already, and will be increasingly, one of the most pressing questions of these critical times.

References

Abimbola, O., Kwesi Aikins, J., Makhesi-Wilkinson, T., & Roberts, E. (2021). *Racism and climate (in)justice: How racism and colonialism shape the climate crisis and climate action*. Heinrich-Böll-Stiftung.

Ajl, M. (2021). *A people's Green New Deal*. Pluto Books.

Almqvist, S., Ullerstam, M., Wärmark, K., Lundberg, D., Lundblad, T., Kanth, M., & Diehl, L. (2020). *Fördjupad analys av den svenska klimatomställningen 2020: Klimat och luft i fokus* [In-depth analysis of the Swedish climate change in 2020: Climate and air in focus]. Naturvårdsvarket. https://www.naturvardsverket.se/Om-Naturvardsverket/Publikationer/ISBN/6900/978-91-620-6945-2/

Anderson, K., Broderick, J. F., & Stoddard, I. (2020). A factor of two: How the mitigation plans of "climate progressive" nations fall far short of Paris-compliant pathways. *Climate Policy, 20*(10), 1290–1304. https://doi.org/10.1080/14693062.2020.1728209

Åström, C., Bjelkmar, P., & Forsberg, B. (2019). Ovanligt många dödsfall i Sverige sommaren 2018—Drygt 600 kan ha dött till följd av värmeböljan [High mortality during the 2018 heat wave in Sweden]. *Lakartidningen*, 116. https://lakartidningen.se/wp-content/uploads/EditorialFiles/FH/%5BFLFH%5D/FLFH.pdf

Åström, D. O., Forsberg, B., Ebi, K. L., & Rocklöv, J. (2013) Attributing mortality from extreme temperatures to climate change in Stockholm, Sweden. *Nature Climate Change, 3*(12), 1050–1054. https://doi:10.1038/nclimate2022

Berglund, O., & Schmidt, D. (2020). *Extinction Rebellion and climate change activism: Breaking the law to change the world*. Springer Nature.

Blühdorn, I. & Deflorian, M. (2019). The collaborative management of sustained unsustainability: On the performance of participatory forms of environmental governance. *Sustainability, 11*, 1189. https://doi.org/10.3390/su11041189

Boda, C. S., & Jerneck, A. (2019). Enabling local adaptation to climate change: Towards collective action in Flagler Beach, Florida, USA. *Climatic Change, 157*, 631–649. https://doi.org/10.1007/s10584-019-02611-6

Boyd, E., Chaffin, B. C., Dorkenoo, K., Jackson, G., Harrington, L., N'guetta, A., & Stuart-Smith, R. (2021). Loss and damage from climate change: A New climate justice agenda. *One Earth, 4*(10), 1365–1370. https://doi.org/10.1016/j.oneear.2021.09.015

Brand, U. & Wissen, M. (2021). *The imperial mode of living: Everyday life and the ecological crisis of capitalism*. Verso.

Brink, E., & Wamsler, C. (2018). Collaborative governance for climate change adaptation: Mapping citizen–municipality interactions. *Environmental Policy and Governance, 28*(2), 82–97. https://doi.org/10.1002/eet.1795

Bruno, K., Karliner, J., & Brotsky, C. (1999). *Greenhouse gangsters vs. climate justice.* Transnational Resource and Action Centre and CorpWatch.

Bullard, R. D., & Wright, B. (Eds.). (2009). *Race, place, and environmental justice after Hurricane Katrina: Struggles to reclaim, rebuild, and revitalize New Orleans and the Gulf Coast.* Routledge.

Burck, J., Hagen, U., Bals, C., Höhne, N., & Nascimento, L. (2021). *Climate change performance index.* Germanwatch, New Climate Institute and Climate Action Network. https://ccpi.org/download/the-climate-change-performance-index-2021/

Carton, W. (2020). Carbon unicorns and fossil futures: Whose emission reduction pathways is the IPCC performing? In J.P. Sapinski, H. Buck, & A. Malm (Eds.), *Has it come to this? The promises and perils of geoengineering on the brink.* Rutgers University Press.

Chatterton, P., Featherstone, D., & Routledge, P. (2013). Articulating climate justice in Copenhagen: Antagonism, the commons, and solidarity. *Antipode, 45*(3), 602–620. https://doi.org/10.1111/j.1467-8330.2012.01025.x

Cheon, A., & Urpelainen, J. (2018). *Activism and the fossil fuel industry.* Routledge.

Ciplet, D., Timmons Roberts, J., & Mizanur, R. K. (2015). *Power in a warming world: The new global politics of climate change and the remaking of environmental inequality.* MIT Press.

Climate Action Lab. (2019). *A people's climate plan for New York City?* The Graduate Center, City University of New York. https://occupyclimatechange.net/wp-content/uploads/2022/03/ClimatePlan_FINAL1-1.pdf

Crow, S. (2014). *Black flags and windmills: Hope, anarchy, and the Common Ground Collective.* PM Press.

Dawson, A. (2017). *Extreme cities: The peril and promise of urban life in the age of climate change.* Verso.

De Moor, J. (2021). Alternative globalities? Climatization processes and the climate movement beyond COPs. *International Politics, 58*(4), 582–599. https://doi.org/10.1057/s41311-020-00222-y

De Moor, J., De Vydt, M., Uba, K., & Wahlström, M. (2021). New kids on the block: Taking stock of the recent cycle of climate activism. *Social Movement Studies, 20*(5), 619–625. https://doi.org/10.1080/14742837.2020.1836617

Derman, B. B. (2020). *Struggles for climate justice: Uneven geographies and the politics of connection.* Springer Nature.

De Rosa, S. P. (2022). Breaking consensus, transforming metabolisms: Notes on direct action against fossil fuels through urban political ecology. *Social Text, 40*(1), 135–155. https://doi.org/10.1215/01642472-9495160

De Rosa, S. P., De Moor, J., & Dabaieh, M. (2022). Vulnerability and activism in urban climate politics: An actor-centered approach to transformational adaptation in Malmö (Sweden). *Cities, 130*, 103848. https://doi.org/10.1016/j.cities.2022.103848

Dietz, M., & Garrelts, H. (Eds.). (2014). *Routledge handbook of the climate change movement*. Routledge.

Dymén, C., & Langlais, R. (2013) Adapting to climate change in Swedish planning practice. *Journal of Planning Education and Research, 33*(1), 108–119. https://doi.org/10.1177/0739456X12463943

Ernstson, H., & Swyngedouw, E. (Eds.). (2018). *Urban political ecology in the Anthropo-obscene: Interruptions and possibilities*. Routledge.

Featherstone, D. (2012). *Solidarity: Hidden histories and geographies of internationalism*. Zed Books.

Gonzalez, C. G. (2020). Racial capitalism, climate justice, and climate displacement. *Oñati Socio-Legal Series, 11*(1), 108–147. https://doi.org/10.35295/osls.iisl/0000-0000-0000-1137

Green, J. F. (2020). Less talk, more walk: Why climate change demands activism in the Academy. *Dædalus, 149*(4), 151–162. https://doi.org/10.1162/daed_a_01824

Haghighatafshar, S., La Cour, Jansen, J. Aspegren, H., Lidström, V., Mattsson, A., & Jönsson, K. (2014). Storm-water management in Malmö and Copenhagen with regard to climate change scenarios. *Vatten: tidskrift för vattenvård/Journal of Water Management and research, 70*(3), 159–168.

Hall, M., Lund, E., & Rummukainen, M. (2015). *Klimatsäkrat Skåne. Centrum för miljö- och klimatforskning*. Lunds universitet.

Han, H., & Ahn, S. W. (2020). Youth mobilization to stop global climate change: Narratives and impact. *Sustainability, 12*(10), 4127. https://doi.org/10.3390/su12104127

Hansen, C. (2020). Alliances, friendships, and alternative structures: Solidarity among radical left activists and precarious migrants in Malmö. *Journal of Race, Ethnicity and the City, 1*(1–2), 67–86. https://doi.org/10.1080/26884674.2020.1797600

Holgersen, S., & Hult, A. (2020). Spatial myopia: Sustainability, urban politics and Malmö City. *International Journal of Urban Sustainable Development, 13*(2), 159–173. https://doi.org/10.1080/19463138.2020.1855432

Holgersen, S., & Malm, A. (2015). "Green fix" as crisis management: Or, in which world is Malmö the world's greenest city? *Geografiska Annaler: Series B, Human Geography, 97*(4), 275–290. https://doi.org/10.1111/geob.12081

Kakenmaster, W. (2019). Articulating resistance: Agonism, radical democracy and climate change activism. *Millennium, 47*(3), 373–397. https://doi.org/10.1177/0305829819839862

Kenis, A. (2021). Clashing tactics, clashing generations: The politics of the school strikes for climate in Belgium. *Politics and Governance, 9*(2), 135–145. https://doi.org/10.17645/pag.v9i2.3869

Kenis, A., & Mathijs, E. (2014). Climate change and post-politics: Repoliticizing the present by imagining the future? *Geoforum, 52*, 148–156. https://doi.org/10.1016/j.geoforum.2014.01.009

Krause, D. (2018). Transformative approaches to address climate change and achieve climate justice. In T. Jafry (Ed.), *Routledge handbook of climate justice* (pp. 509–520). Routledge.

Kurtz, R. M. (2020). Direct action and the climate crisis: Interventions to resist and reorganize the metabolic relations of capitalism. *Radical Philosophy Review, 23*(2), 261–297. https://doi.org/10.5840/radphil-rev2020813114

Machin, A. (2013). *Negotiating climate change: Radical democracy and the illusion of consensus*. Zed Books.

Martinez-Alier, J. (2002). *The environmentalism of the poor: A study of ecological conflicts and valuation*. Edward Elgar Publishing.

Méndez, M. (2020). *Climate change from the streets: How conflict and collaboration strengthen the environmental justice movement*. Yale University Press.

Mouffe, C. (2000). *The democratic paradox*. Verso.

Newell, P., Srivastava, S., Naess, L. O., Torres Contreras, G. A., & Price, R. (2021). Toward transformative climate justice: An emerging research agenda. *Wiley Interdisciplinary Reviews: Climate Change*, 12(6). https://doi.org/10.1002/wcc.733

Nylund, K. (2014). Conceptions of justice in the planning of the new urban landscape: Recent changes in the comprehensive planning discourse in Malmö, Sweden. *Planning Theory & Practice, 15*(1), 41–61. https://doi.org/10.1080/14649357.2013.866263

Perry, K. (2021, July 27). (Un)just transitions and black dispossession: The disposability of Caribbean "refugees" and the political economy of climate justice. *Politics*. https://ssrn.com/abstract=3900602

Rancière, J. (2010). *Dissensus: On politics and aesthetics*. Bloomsbury.

Red Nation. (2021). *The red deal: Indigenous action to save our Earth*. Common Notions.

Rice, J., Long, J., & Levenda, A. (2021). Against climate apartheid: Confronting the persistent legacies of expendability for climate justice. *Environment and Planning E: Nature and Space, 5*(2), 625–645. https://doi.org/10.1177/2514848621999286

Rósen, R. (2020, January 15). Nu råder det klimatnödläge i Malmö [There is now a climate emergency in Malmö]. *Aftonbladet*. https://www.aftonbladet.se/nyheter/a/rAKLww/nu-rader-det-klimatnodlage-i-malmo

SCCA [Swedish Civil Contingencies Agency]. (2018). *If crisis or war comes*. https://www.dinsakerhet.se/siteassets/dinsakerhet.se/broschyren-om-krisen-eller-kriget-kommer/om-krisen-eller-kriget-kommer---engelska-2.pdf

SCCV [Swedish Commission on Climate and Vulnerability]. (2007). *Sweden facing climate change—Threats and opportunities*. Swedish Government Official Reports.

Schlosberg, D., & Collins, L. B. (2014). From environmental to climate justice: Climate change and the discourse of environmental justice. *Wiley Interdisciplinary Reviews: Climate Change, 5*(3), 359–374. https://doi.org/10.1002/wcc.275

Schmelzer, M., Vetter, A., & Vansintjan, A. (2022). *The future is degrowth: A guide to a world beyond capitalism*. Verso.

SCPC [Swedish Climate Policy Council]. (2021). *Annual Report n. 4*. Stockholm.

Sikor, T., & Newell, P. (2014). Globalizing environmental justice? *Geoforum, 54*, 151–157. https://doi.org/10.1016/j.geoforum.2014.04.009

SKL. (2019). *Klimatförändringarnas lokala effekter. Exempel från tre kommuner* [Local effects of climate change: Examples from three municipalities]. Sveriges Kommuner och Landsting [Sweden's municipalities and county councils], Stockholm.

Stoddard, I., Anderson, K., Capstick, S., Carton, W., Depledge, J., Facer, K., et al. (2021). Three decades of climate mitigation: Why haven't we bent the global emissions curve? *Annual Review of Environment and Resources, 46*, 653–689. https://doi.org/10.1146/annurev-environ-012220-011104

Swyngedouw, E. (2010). Apocalypse forever? *Theory, culture & society, 27*(2–3), 213–232. https://doi.org/10.1177/0263276409358728

Swyngedouw, E. (2011). Depoliticized environments: The end of nature, climate change and the post-political condition. *Royal Institute of Philosophy Supplement 69*, 253–74. https://doi.org/10.1017/S1358246111000300

Swyngedouw, E. (2014). Depoliticization ("the political"). In G. D'Alisa, F. Demaria, & G. Kallis (Eds.), *Degrowth: A vocabulary for a new era*. Routledge.

Temper, L., Avila, S., Del Bene, D., Gobby, J., Kosoy, N., Le Billon, P., & Walter, M. (2020). Movements shaping climate futures: A systematic mapping of protests against fossil fuel and low-carbon energy projects. *Environmental Research Letters, 15*(12), 123004. https://doi.org/10.1088/1748-9326/abc197

Vandepitte, E., Vandermoere, F., & Hustinx, L. (2019). Civil anarchizing for the common good: Culturally patterned politics of legitimacy in the climate justice movement. *VOLUNTAS: International Journal of Voluntary and Nonprofit Organizations, 30*(2), 327–341. https://doi.org/10.1007/s11266-018-00073-5

Walker, G. (2012). *Environmental justice: Concepts, evidence and politics*. Routledge.

Watt, R. (2021). The fantasy of carbon offsetting. *Environmental Politics, 30*(7), 1069–1088. https://doi.org/10.1080/09644016.2021.1877063

XR Sweden. (2019). *When the climate crisis comes*. https://www.extinctionrebellion.se/wp-content/uploads/2019/11/Klimatkrisen-eng.pdf

Yang, W., Gardelin, M., Olsson, J., & Bosshard, T. (2015). Multi-variable bias correction: Application of forest fire risk in present and future climate in Sweden. *Natural Hazards and Earth System Sciences*, 15(9), 2037–2057. https://doi.org/10.5194/nhess-15-2037-2015

Interviews

A., FGF, April 28, 2020.
J., XR Skåne, March 27, 2019.
L., KNT, November 13, 2019.
M., KNT, November 13, 2019.
M., XR Skåne, October 10, 2019.

Speeches

K., FGF, February 11, 2018.

About the author

Salvatore Paolo De Rosa is a researcher at the Center for Applied Ecological Thinking, University of Copenhagen. Previously, he was a researcher at Lund University in Sweden, where he also received his PhD in human geography. With a background in anthropology and political ecology, his research focuses on collective action and socio-environmental transformations towards sustainability and justice.

5. Catalyzing transformational action for climate change adaptation

The Ala Wai management plan in Honolulu, Hawai'i, USA

Valentine Huet

Abstract

This chapter deals with adaptation challenges and governance complexities in the context of coastal cities and islands, examining the different views and strategies concerning the Ala Wai Flood Risk Management Project in Honolulu, the economic and political center of Hawai'i. It focuses on the political tensions entangling Hawai'ian residents, activists, politicians and the military engineers involved in the project, detailing the clash between the economic growth and the environmental justice paradigms. Drawing from indigenous Hawai'ian knowledge, deeply rooted in the history and ecology of the islands, Huet proposes an hybrid paradigm shift through a transformation strategy based on the indigenous Hawai'ian practice of *ho'oponopono*: a healing process to repair past injustices and bring reconciliation through a problem-solving attitude.

Keywords: indigenous knowledge, development, flood risk, decolonization, Hawai'i

The 21st century is marked by the threat of climate change, and cities are at the core of this global challenge (Rosenzweig et al., 2018). Coastal cities in particular are on the front lines. Hawai'i, in the Pacific region, is no exception and is already experiencing significant changes (NCA, 2014). In the *Third National Climate Assessment: Climate Change Impacts in the United States* (ibid.), one of the identified threats for the Pacific archipelago is sea level rise coupled with flood hazards. In 2017, the National Oceanic and Atmospheric Administration (NOAA), the US government's climate

---

Armiero, M., E. Turhan & S. Paolo De Rosa (eds), *Urban Movements and Climate Change: Loss, Damage and Radical Adaptation*. Taylor & Francis Group 2023
DOI: 10.5117/9789463726665_CH05

research center, registered sea level anomalies of up to 40 centimeters in Hawai'i and also detected that sea surface temperatures rose about 2 degrees and recorded an increase in extreme rainfall with 100% more precipitation than the usual average (Kruk, 2017).

All of these risks threaten infrastructures and, more generally, urban governance. Honolulu, on the island of O'ahu, the economic and political center of Hawai'i, is facing double threats. On the one hand, sea level rise threatens the Ala Wai Canal (*ala wai* literally means "waterway" in Hawai'ian) and thus Waikiki, the heart of O'ahu's tourism. On the other hand, it is at risk of flash floods from the mountains, which are barely retained by the Ala Wai. After its annexation of the island nation, the US government built the Ala Wai Canal in 1921 to control the water flow through the city. Due to climate threats, the Ala Wai is currently subject to a flood risk adaptation management plan supervised by the US Army Corps of Engineers (USACE). Grassroots movements such as Protect Our Ala Wai Watershed (POAWW) have expressed their concerns about the social and sustainability implications of this government project.

This chapter examines the different views and strategies concerning the Ala Wai Flood Risk Management Project as expressed by Hawai'i residents, activists, politicians and the military engineers involved in the project. It is guided by these research questions:

- Why are some Hawai'ians in the POAWW grassroots movement challenging a plan that is supposed to be a form of climate change adaptation?
- How can a transformation strategy help grassroots movements, governments, planners and engineers to create sustainable and inclusive urban projects in the context of climate change?

In what follows, I explore the clash between the logic of the state, based mainly on economic viability, and indigenous Hawai'ian knowledge, which is deeply rooted in the history and ecology of the islands.

The fieldwork for this chapter took place in Honolulu from January to the beginning of March 2020 as part of the Occupy Climate Change! project. The main methodology followed was grounded research coupled with an ethnographic approach. Participant observation and in-depth and semi-structured interviews were conducted to understand the different points of view on the conflict within Hawai'ian society. Archival research included examining old photos of Ala Wai in the Hawai'i State Archives and articles from the local newspaper, *The Star Alliance*. This research helped me to interpret the historical aspects of the infrastructure and Hawai'ian history and thus understand the evolution of the conflict over the Ala Wai between Hawai'ian communities and authorities before and after the American presence in Hawai'i.

Some methodological, analytical and cultural limits affected the field work. The first limit was due to the researcher's positionality as a European. Indeed, the analysis was done through the lens of a white European woman who had not experienced inequalities due to indigenousness before embarking on this field study. Second, being an outsider to Hawai'i caused difficulties with access and doing the field work, including skepticism on the part of the grassroots movements, citizens and state representatives. Finally, because of complex historical relations between Europeans coming to Hawai'i and the continuous fear of being betrayed, the native Hawai'ians are used to protecting their culture by limiting outsiders' entry into their groups. Despite these limitations, this case study offers some useful insights toward a hybrid paradigm on climate change adaptation, including paying attention to the role of decolonizing healing.

Case study: The Ala Wai Canal

The Ala Wai Canal lies in the heart of Hawai'i's indigenous historical farming zone. The canal encompasses three *ahupua'a*: Palolo, Manoa (which includes Moiliili and Waikiki) and Makiki. An *ahupua'a* is a slice of the island, which was traditionally delimited from the top of the local mountain to the shore (Minerbi, 1999). The *ahupua'a* management system, used since the early settlers came to Hawai'i around 600 AD, was a unique land-use subdivision for its interlinkages and complementarity based on an exchange system free from monetary transactions (Beamer, 2014). Divided into three zones, coastal, agricultural and forest, it was an adaptive model for the specific climate and geography of the islands (ibid.). This way of dividing up the land ensured the success of the three zones by complementing and sustaining each other's resources (ibid.). Multiple streams from the three *ahupua'as* feed the plain of Waikiki, which was once a vast marshland.

However, after the occupation and colonization of Hawai'i in 1898, the American government imposed a new type of land-use management intended to increase the profitability of the land. The canal was an example of this new land-use management based on economic, rather than socio-ecological, viability. The Ala Wai Canal was commissioned in 1921 by Lucius E. Pinkham, then governor of the Territory of Hawaii, and was completed in 1928. The control of water flows radically changed Honolulu's watershed.

Figure 5.1. Waikiki (on the left) facing Moiliili (on the right), with the Ala Wai Canal between them. Photo: Darly Mitchell, CC BY-SA 2.0.

It merged the existing streams into centralized waterways to get water to flow into the Ala Wai. The artificial canal serves as Honolulu's northern border in Waikiki. One consequence of the rapid and massive urbanization of Waikiki is that now the impervious surfaces are unable to absorb water and thus increase the risk of both flooding and sea level rise (Connelly, 2020). Over the years, the canal has also suffered from erosion issues, and multiple improvements have been made.

The development of the Ala Wai had socioeconomic in addition to environmental consequences (Okamoto, 2016). On the one hand, Waikiki is an the economically powerful district, which generates 8% of Hawai'i's gross domestic product thanks to tourism (Hawai'i Department of Business, Economic Development & Tourism, 2003). On the other hand, Hawai'ians suffer from vast disparities of income. The surroundings of Ala Wai are representative of these inequalities. According to Hawai'i's 2015 Asset Limited, Income, Constrained, Employed (ALICE) Report, 48% of the households in Hawai'i live below the median income level. This figure is higher for native Hawai'ians (57%), of which 10% live on less than $15,000 per year per single adult (2015, p. 19). The Ala Wai area has a particular socioeconomic and land ownership structure. Its socioeconomic disparities are mixed between poverty and high income, depending on the *ahupua'a*. After annexation by the United States, most of the native Hawai'ians were pushed out of central Honolulu toward Moiliili. The Moiliili area, a relatively impoverished area, faces wealthy Waikiki, but is separated physically by the Ala Wai. Figure 5.1 shows this contrast clearly with the hotels and condominiums of Waikiki on the left side of the photo and single-story homes in Moiliili in the background on the right side. If the Ala Wai Flood Risk Management Project becomes reality, a massive floodwall designed to preserve Waikiki's economy could come at the expense of Moiliili.

The Ala Wai Flood Risk Management Project

The Ala Wai Flood Risk Management Project is a prime example of a climate change adaptation (CCA) plan. Hawai'i's Department of Land and Natural Resources (DLNR) started the Ala Wai restoration project in 1994. Four years later, it asked USACE to take over the project by expanding it to include flood control. From 1998 to 2011, they both worked jointly with a variety of local community groups. However, in 2011, USACE had to come up with a plan or the project's funding would be withdrawn. Eventually, in July 2018, Congress allocated $345 million to complete the proposed plan, for which nonfederal institutions (state and city council) were obligated to provide $125 million.

The management plan includes 11 projects in one and encompasses seven districts. The funded project plan includes the development of 11 structural components of one main structure, which stands alone for the in-stream debris catchment. It also includes four retention basins, five debris/retention basins located on the upper stream, and the floodwalls along the Ala Wai consisting of two pump stations for the canal's internal drainage (USACE, 2020). USACE justifies the project by claiming that it will protect infrastructure, the economy, as well as people from potential flooding events such as a 100-year storm (ibid.). USACE projects that such flooding could cost the Hawai'ian state $1.14 billion in structural damage, and that it would affect 54,000 inhabitants and 80,000 tourists (Okamoto, 2016). Now that the plan has been funded, it is hard to modify it and include alternative solutions to the one proposed by USACE.

In accordance with Hawai'ian land-use management law, USACE created an environmental impact statement (EIS) that it submitted in 2017. According to Hawai'ian law, creating an EIS is a process that should involve all concerned stakeholders and their points of view before starting the design and planning of a project. However, USACE completed the EIS without including all the potentially affected stakeholders. The POAWW grassroots movement sued USACE and won the case since the EIS was created improperly. During interviews, some interviewees stated that the decision to sue USACE was part of a larger process of doing justice to local inhabitants, but enabled POAWW to have enough time to convince USACE to include local and nature-based solutions (NBSs) better suited for the context of Hawai'i in its plan. Since the judgment, USACE has held open meetings with civil society organizations to complete the improperly filed EIS with smaller retention basins. However, USACE engineers are still not considering the inclusion of NBSs, despite the people's insistence.

The Protect Our Ala Wai Watershed (POAWW) grassroots movement

The POAWW grassroots movement is a nonprofit organization that was founded in 2019. It includes residents of the three *ahupua'as* and carries out local actions to increase public and political awareness of the ongoing conflict over the Ala Wai flood risk adaptation management plan. First, the movement objected to USACE's failure to disseminate information to civil society and its lack of transparency. Second, it denounced the current plan's protecting of Waikiki without eliminating threats to other *ahupua'as*. Indeed, some activists calculated that the planned massive walls next to the canal would flood Moiliili. Likewise, the upstream retention basins are planned to retain water so that floodwaters do not end up in Waikiki, but flood upstream areas instead. Third, POAWW criticizes the absence of NBSs in the proposed plan and its undervaluation of natural resources. NBSs are embedded in the thinking of ecosystem services (Keestra et al., 2018, p. 999), which would be, according to POAWW, better suited for the local environment. To be successful, an NBS has to integrate the local environmental understanding of a place. NBSs have shown excellent results for river protection and coastal defense (ibid.).

Sean Connelly, a member of POAWW and a Hawai'ian architect, promotes NBSs inspired by indigenous Hawai'ian knowledge as part of an alternative for the Ala Wai Flood Risk Management Project. The POAWW grassroots movement emphasizes the importance of incorporating indigenous Hawai'ian knowledge into the USACE plan. The architect has proposed alternative solutions to the management plan, which are available on his website, Ala Wai Centennial. His work uses a social and ecological approach to answer the technical challenges—solutions such as the restoration of the marshland in the golf course next to Ala Wai, transforming it into a public space, as well as the use of *lo'i kalos* (taro patches) on the upper streams to retain water. These would restore the ecosystem of the *ahupua'a* and its ability to retain water while reducing the flooding of poorer areas (POAWW, 2020). According to Connelly, *ahupua'a* recovery could be a new sustainable door to the future of Hawai'i's economy and also address the island's social justice and climate change issues. Recognition of the necessity of including indigenous Hawai'ian knowledge has been spreading in Hawai'i for years now along with a global indigenous resurgence movement.

Climate change adaptation: Processes and paradigms

Adaptation to climate change is an unavoidable task, given the urgency of the climate crisis (IPCC, 2018). Adaptation is defined by the IPCC as the "process of adjustments to actual or expected climate and its effects" (IPCC, 2014, p. 18). As such, the climate change adaptation (CCA) process is a dynamic process with complex interlinkages in response to climate change, which is inherent to human lives (Pelling, 2011). The plurality of responses and actions initiated for the CCA process are represented in different narratives in different societies (Watts, 1983).

More and more, cities are producing CCA strategies to respond to climate change threats (Pelling, 2011). These multiple strategies differ depending on governments' social, political and economic agendas (Bulkeley & Betsill, 2005). There are currently two dominant urban planning trends linked with CCA (Pelling, 2011): the resilience and the transition strategies. Pelling's work (ibid.) describes resilience strategies as maintaining the status quo (p. 3) and transition strategies as bringing incremental change (ibid.). These two trends are opposed since they involve different actions and interests that are inherent to two different paradigms: on the one hand, the economic growth paradigm for the resilience strategies, and, on the other hand, the environmental justice paradigm for the transition strategies. To surpass these two paradigms, here I propose a hybrid paradigm with roots in the transformational strategies that produce new postulates with a nonbinary vision.

The economic growth paradigm

The economic growth paradigm feeds into resilience strategies, which are part of the dominant discourse worldwide (Laberenne & Lamson Hall, 2018).

The core of these strategies is the need to quantify the objectives and the results of CCA solutions to decide whether a solution works (ibid.). Resilience strategies articulate adaptation processes to potential shocks through the economic and environmental lenses (Thieme, 2016). In two ways, urban resilience strategies tend to diversify the weak points of the system and try to spread risks to limit their consequences (ibid.). First, protectionism of what has already been built in the name of economic growth is associated with resilience strategies (Pelling, 2011). Second, the expansion of the urban environment is based on the modernist vision of land-use management that fosters its profitability to meet to the city's future needs while creating dependence on Western knowledge, culture and capitalist markets. Thus, resilience strategies are, in most cases, allied with the capitalist system and its perpetuation (Webber, 2016, p. 408).

As a consequence of growth-fueled capitalism, the current CCA processes in modern societies foster social inequalities because they are predicated on a top-down approach and increase disparities in resources (Pelling, 2011). Morchain's work (2018) demonstrated that most CCA processes do not consider local values and traditions. When they do so, they attempt to accommodate the country's geopolitical and economic interests (ibid., p. 68) and their outcomes tend not to empower people but, on the contrary, to maintain their marginalization (ibid.). This way of thinking is described as informative and realistic (Nyamwanza & Bhatasara, 2014) and does not consider the social constructions of societies as key factors in decisions, but the causality chains of more tangible elements. These tendencies show that resilience strategies are affiliated with the post-positivist approach (Guba & Lincoln, 1994).

The environmental justice paradigm

The environmental justice paradigm is rooted in the social, economic and environmental injustices and mismanagement of the economic-growth paradigm. The environmental justice paradigm seeks to advocate for fair treatment and rejects the discrimination of certain communities because of their race or income (Daly, 1996). In response to the economic growth paradigm's top-down governance approach, the environmental justice paradigm highlights the need for a bottom-up approach (Gunnersson & Svensfelt, 2017). This governance approach's paradigm fosters a co-construction of knowledge and capacities among all citizens (Daly, 1996). Thus, it incentivizes people to take control of their local environment so that it can be rendered coherent with local values (Gunnersson & Svensfelt, 2017). Moreover, local knowledge produced by people and by cultural processes is seen as a unique way of understanding and using ecological resources (Leonard et al., 2013).

That is why the environmental justice paradigm mainly occurs at local scales through practices anchored in ecological knowledge of the environment and stimulation of local forms of production rather than in large-scale interventions, as the economic growth paradigm tends to do (Hahnel, 2007). The environmental justice paradigm advances the need for social and cognitive knowledge and seeks to build trust between social groups (Carr, 2004). It also demands the recognition and participation of all citizens in the decision-making system (Agyeman et al., 2016). These components indicate that the environmental justice paradigm uses a social-constructivist approach (Adamson et al., 2002, p. 115), leading to incremental changes and thus, the transition strategy as described by Pelling (2011).

Understanding the value systems of paradigms

The two paradigms described above are embedded in different value systems that deeply influence the choice of a certain strategy by a society. Indeed, the different types of understandings and actions of societies, caused by different worldviews, are rooted in different value systems (Adger et al., 2009). The value system described by Williams (1979) is an organized set of preferential standards that are used for the organization and construction of a given society. The link between people's motivation and individual values is created by an organization of coherent structures in a system of values (O'Brien & Wolf, 2010). These worldviews can change significantly over time and between generations or even within an individual's lifespan (ibid.).

O'Brien and Wolf (2010) studied the value system of societies in the context of climate change. They stress that, when it comes to transition and vulnerability due to climate change, the primary stake is about how people perceive it. This perception, transformed into values, determines their actions (ibid.). O'Brien and Wolf claim that CCA plans should consider and include local value systems and use an integrated approach to be sustainable in the short and long term (ibid., p. 239). In the context of Ala Wai, the clash between the economic growth and environmental justice paradigms leads to failure, while indicating the need to attempt some kind of progressive compromise that might advance the purposes of the latter, although within the mainstream framework of the former. In other words, while radically subverting the current mainstream paradigm, it may be still possible to foster a transformative agenda within it.

Towards autonomy

This case study highlights profound divergences of the value systems and solutions' approaches of the different stakeholders for the same climate change adaptation project. On the one hand, USACE is promoting the resilience strategy based on the economic growth paradigm by reinforcing the competitiveness of Waikiki and functioning of the Ala Wai. USACE's values and actions aim toward continuous and productive economic growth. They promote protectionism of what has been created—in this case, the Ala Wai Canal and Waikiki. USACE developed a top-down management approach since the beginning of the plan with the improperly completed EIS, which led to a weak understanding of social and environmental dynamics, needs and values.

On the other hand, POAWW, whose narrative is based on local indigenous knowledge, embodies the transition strategy. This strategy has an in-depth knowledge of the local environment and an extensive understanding of how to use natural and local resources. The use of indigenous Hawaiʻian knowledge is reflected in this strategy. Community is essential to the grassroots movement's dynamic since social networks are at the heart of the decision-making and of this transition strategy. Environmental and social injustices are the motives of POAWW's actions, which emphasize a bottom-up governance approach. These two strategies oppose one another in many ways and are hard to reconcile. They appear to be two separate paths based on two different understandings of the world.

The need for a paradigm shift lies in part in climate-resilient development, an idea developed by the most recent IPCC report on impacts, adaptation and vulnerability. The IPCC (2022) defines climate-resilient development as a "solutions framework" that combines strategies for dealing with climate risks with efforts to reduce greenhouse gas emissions, resulting in improvements in the well-being of people and the natural environment. The notion of climate-resilient development fosters the need for a paradigm shift between engineering solutions toward more adequate local solutions with multiple voices in decision-making processes. According to the United Nations Framework Convention on Climate Change (UNFCCC), climate-resilient development is defined by multiple trajectories that would avoid "dangerous anthropogenic interference with the climate system" (United Nations, 1992, p. 4). In order to reduce human impacts, the UNFCCC's notion of climate-resilient development raises the necessity of transformational change. Transformational change is an ongoing process that manages changes by encompassing the different driving forces that affect sustainable development. Transformational change is a combination of innovativeness, flexibility and participative problem-solving (IPCC, 2018). Its goals are to mitigate and adapt to climate change (ibid.). This new type of change has to be infused into all the bodies of society—political, societal and economic—where some values could be different (ibid.). Therefore, transformational change implies new types of thinking and actions in the public sphere.

To address such complex issues, Arturo Escobar (2018) offers a "pluriverse design approach," which is an inclusive type of thinking, especially toward societies' multiple value systems. His theoretical approach promotes individual autonomy to create real changes beyond the capitalist and central state, thanks to everyone's participation (ibid.). Hence, autonomy, as an inclusive notion, puts indigenous values and knowledge, and the economic-growth understandings on the same level. Furthermore, Gustavo Esteva (2015) also proposes that autonomy acknowledges cultural and political practices as much as economic ones. It oscillates between two types of politics: "the politics of the real and the politics of the possible" (Escobar, 2018, p. 226). The "politics of the real" is a reference to today's modern Western politics, while the "politics of the possible" echoes a new type of politics grounded in autonomy and creativity. Indeed, to surpass the current cleavages of the "politics of the real," people need to be emancipated individually, but even more so collectively, from the current political framework to create their future of the "possible." The choice of the implementation of a certain type of politics depends on collective values and actions which arise in the practical, political and personal spheres as described by Adger et al. (2009).

In the Ala Wai Flood Risk Management Project, the clash between the economic growth and the environmental justice paradigms is evident. No common ground appears to be available since USACE has the funding for specific adaptation projects and the final say concerning the project. Not having negotiations serves the interests of USACE, so it can proceed with the management plan at a faster pace. However, by acting in the public sphere and showing their disagreement, members of the grassroots movement asks for acknowledgment of their ideas and solutions. Consequently, even USACE may find it useful to look for some sort of compromise—perhaps what Belayneh (2017) has called "collective paradigm shifts"—that could settle the differences.

Propositions for a hybrid paradigm shift

In the context of the Ala Wai Flood Risk Management Project, the shift toward the hybrid paradigm would transform the entire project and bear deep implications for the different stakeholders. The involvement of all the stakeholders in the project development process is necessary to represent the value systems of the people living on the island. This implies that all stakeholders should be able to propose a list of solutions and justify them. After discussions, these solutions would then be voted on by the inhabitants of O'ahu. This democratic process is essential so that everyone living on the island can participate in the final plan, for social and spatial justice issues affect everyone in O'ahu. This would allow everyone concerned to be part of the conversation and have a say. This will make decisions be the result of a plurality and representative of the dynamics of the different value systems on the island. The solutions will have to be incorporated into the project and implemented by USACE. This inclusive approach would allow for including creative ideas and projects to meet the requirements of the flood mitigation and adaptation process.

The nonfinite temporal character of transformation strategies brings another dimension to the CCA project. Indeed, as infrastructure projects, CCAs are subjected to constant social and environmental change. The Ala Wai Flood Risk Management Project would therefore not be a definite project. Instead, it would evolve with time and future needs. This unique temporal scale should be considered a constant adaptation process. As such, it will be able to evolve depending on the future hazards and social needs of the island. Due to the multiparty governance described, every citizen will be the guarantor of this variable aspect and will have the right to denounce nonaccordance or restrictions to the Environmental Court of Hawai'i, which will be the highest guarantor of the Ala Wai. The transformation strategy suggests that the evolution would not only be subjected to economic rents. Thus, different factors such as social, cultural and environmental necessities will be added to the decision-making process. They will also be the criteria that the Environmental Court will have to consider while judging potential abuses of the Ala Wai. This protection is to guarantee that no militarization or economic growth processes will take precedence over environmental or social matters.

Further steps toward a decolonizing healing process

In the case of Hawai'i, an acknowledgment and active use of indigenous Hawai'ian knowledge by the state seems necessary in order to surpass the current rifts. It would include, in terms of the project development, spiritual and cultural practices that matter to the Hawai'ian community. Healing processes are central to indigenous Pacific cultures. These processes inspire many of their actions with the idea of settling a constructive, healthy and sustainable basis for the future. Though acknowledging my positionality as a white European outsider, I would humbly explore the revolutionary potentialities of the spiritual and indigenous Hawai'ian practice of *ho'oponopono*. *Ho'oponopono* is defined in the Hawai'ian dictionary as: "To put to rights; to put in order or shape, correct, revise, adjust, amend, regulate, arrange, rectify, tidy up, make orderly or neat, administer, superintend, supervise, manage, edit, work carefully or neatly; to make ready, as canoe men preparing to catch a wave" (Wehe Wehe, 2019). This healing process repairs past injustices through forgiveness and brings repentance and reconciliation through a problem-solving attitude (Shock, 1985). In the case of the Ala Wai Flood Risk Management Project, *ho'oponopono* could help to restart the project development as a process of healing and reconciliation that acknowledges injustices.

Traditional *ho'oponopono* is led by a *kahuna lā'au lapa'au* (healing priest) to heal physical or mental illnesses within a family. *Ho'oponopono* corrects deviant behavior, restores good relationships between family members and maintains them (Parsons, 1995). During the process, thanks to the gods and their respective powers and affiliations, they reach the causes of conflicts (ibid.). Usually, the elders of the family direct spiritual purification. If the family cannot resolve the dispute, they turn to a respected and qualified third party.

In the context of this case, the tradition could be developed using this metaphor: the "deviant" behavior could be USACE failing to address the CCA processes from the point of view shared by Hawai'ian civil society, which would be considered "the family." "The family" will try to resolve this conflict by sharing their deliberations on the situation. This implies that indigenous Hawai'ian knowledge will overrule Western thinking. Thanks to its reciprocity of acknowledgment, this arrangement may create a transfer of knowledge and practices leading toward transformational changes and fostering the autonomy of the collective. In a decolonizing perspective, this practice could bring reciprocity and develop a plan for the Ala Wai made out of local solutions and a close collaborative process with all the citizens who wish to be part of it.

Conclusion

This chapter addressed climate change adaptation in terms of a radical clash between the economic growth and the environmental justice paradigms. This clash is even stronger in the case of colonial societies where the Western framing of a technical problem is considered superior to local indigenous knowledge about the topic. Without renouncing the objective of subverting the present unjust socio-ecological system, this chapter has explored the possibilities of fostering transformational actions that can create a space in between where the claims of the environmental justice approach might find a way through the mainstream economic growth paradigm. Besides exploring possible compromises, I have made an attempt to envision a different path. By respectfully invoking the indigenous Hawai'ian practice of *ho'oponopono*, decision-makers and local communities can subvert the hierarchies of the legitimation of knowledge production and social innovations for adaptation while acknowledging past wrong-doings and injustices in any quest for healing and reconciliation. The possibility of fostering meaningful climate change adaptation lies not in shallow compromise between divergent value systems, but in nurturing radical healing and decolonizing reconciliation processes.

References

Adamson, J., Evans, M., & Stein, R. (2002). *The environmental justice reader: Politics, poetics, and pedagogy*. University of Arizona Press.

Adger, W. N., Irene Lorenzoni, I., & O'Brien, K. L. (Eds.). (2009). *Adapting to climate change: Thresholds, values, governance*. Cambridge University Press.

Agyeman, J., Schlosberg, D., Craven, L., & Matthews, C. (2016). Trends and directions in environmental justice: From inequity to everyday life, community, and just sustainabilities. *Annual Review of Environment and Resources, 41*, 321–340. https://doi.org/10.1146/annurev-environ-110615-090052

Aloha United Way. (2015). *United Way Alice Report—Hawai'i*. https://labor.hawaii.gov/wdc/files/2018/12/ALICE-report.pdf

Beamer, K. (2014). *No mākou ka mana: Liberating the nation*. Kamehameha Publishing.

Belayneh, A. (2017). Paradigm shifts: Climate change and water allocation. *Oceanography & Fisheries, 5*(1). https://juniperpublishers.com/ofoaj/pdf/OFOAJ.MS.ID.555652.pdf

Bulkeley, H., & Betsill, M. (2005). Rethinking sustainable cities: Multilevel governance and the "urban" politics of climate Change. *Environmental Politics, 14*(1), 42–63. https://doi.org/10.1080/0964401042000310178

Carr, M. (2004). *Bioregionalism and civil society: Democratic challenges to incorporate to globalism.* UBC Press.

Connelly, S. (2020). Ala Wai Centennial. https://www.alawaicentennial.org

Daly, H. E. (1996). *Beyond growth: The economics of sustainable development.* Beacon Press.

Dujardin, S. (2019). Planning with climate change? A poststructuralist approach to climate change adaptation. *Annals of the American Association of Geographers, 110*(4), 1059–1074. https://doi.org/10.1080/24694452.2019.1664888

Escobar, A. (2018). *Designs for the pluriverse: Radical interdependence, autonomy, and the making of worlds.* Duke University Press.

Esteva, G. (2015). The hour of autonomy. *Latin American and Caribbean Ethnic Studies, 10*(1), 134–145. https://doi.org/10.1080/17442222.2015.1034436

Guba, E. G., & Lincoln, Y. S. (1994). Competing paradigms in qualitative research. In N. K. Denzin & Y. S. Lincoln (Eds.), *Handbook of qualitative research* (pp. 105–117). Sage.

Gunnersson, U., & Svensfelt, A. (2017). Sustainability discourses and justice towards social-ecological justice. In R. Holifield, J. Chakraborty, & G. Walker (Eds.), *The Routledge handbook of environmental justice* (pp. 160–171). Routledge.

Hahnel, R. (2007). Eco-localism: A constructive critique. *Capitalism Nature Socialism, 18*(2), 62–78. https://doi.org/10.1080/10455750701366444

Hawai'i Department of Business, Economic Development & Tourism. (2003). The economic contribution of Waikiki. http://files.hawaii.gov/dbedt/economic/data_reports/e-reports/econ_waikiki.pdf

Hulme, M. (2009). *Why we disagree about climate change.* Cambridge University Press.

IPCC. (2014). Annex II: Glossary [Mach, K. J., S. Planton and C. von Stechow (Eds.)]. In Core Writing Team, R. K. Pachauri, & L. A. Meyer (Eds.), *Climate change 2014: Synthesis report—Contribution of Working Groups I, II and III to the Fifth Assessment Report of the Intergovernmental Panel on Climate Change* (pp. 117–130).

IPCC. (2018). Climate-resilient pathways: Adaptation, mitigation, and sustainable development. https://www.ipcc.ch/site/assets/uploads/2018/02/WGIIAR5-Chap20_FINAL.pdf

IPCC. (2022). FAQ 6: What is climate resilient development and how do we pursue it? https://www.ipcc.ch/report/ar6/wg2/about/frequently-asked-questions/keyfaq6/

Keestra, S., Nunes, J., Novara A., Finger, D., Avelar, D., Kalantari, Z., & Cerdà, A. (2018). The superior effect of nature-based solutions in land management for enhancing ecosystem services. *Science of the Total Environment, 610–611*, 997–1009. https://www.doi.org/10.1016/j.scitotenv.2017.08.077

Kruk, M. (2017, December 26). In Hawaii and the Pacific Islands, 2017 has been anything but normal. *Beyond the Data.* https://www.climate.gov/news-features/blogs/beyond-data/hawaii-andpacific-islands-2017-has-been-anything-normal

Laberenne, R., & Lamson Hall, P. (2018). Planning for urban growth for a more resilient future. *100 Resilient Cities.* https://www.100resilientcities.org/planning-urban-growthresilient-future/

Leonard, S., Parsons, M., Olawsky, K., & Kofod, F. (2013). The role of culture and traditional knowledge in climate change adaptation: Insights from East Kimberley, Australia. *Global Environmental Change, 23*(3), 623–632. https://doi.org/10.1016/j.gloenvcha.2013.02.012

Minerbi, L. (1999). Indigenous management models and protection of the *Ahupua'a. Social Process in Hawai'i, 39*, 208–225.

Morchain, D. (2018). Rethinking the framing of climate change adaptation. In S. Klepp and L. Chavez-Rodriguez (Eds.), *A critical approach to climate change adaptation: Discourses, policies, and practices.* Routledge.

NCA. (2014). *Third national climate assessment: Climate change impacts in the United States.* https://nca2014.globalchange.gov/highlights/regions/hawaii#statement-16782

Nyamwanza, A., & Bhatasara, S. (2014). The utility of postmodern thinking in climate adaptation research. *Environment Development and Sustainability, 17*(5), 1183–1196. https://doi.org/10.1007/s10668-014-9599-5

O'Brien, K. (2018). Is the 1.5°C target possible? Exploring the three spheres of transformation. *Current Opinion in Environmental Sustainability, 31*, 153–160. https://doi.org/10.1016/j.cosust.2018.04.010

O'Brien, K. L., & Wolf, J. (2010). A values-based approach to vulnerability and adaptation to climate change. *WIREs Climate Change, 1*, 232–242. https://doi.org/10.1002/wcc.30

Okamoto, K. (2016, May 1). Honolulu might be sunk due to climate change. *Newsweek.* http://www.newsweek.com/2016/05/13/hawaii-ala-wai-canalproject-waikiki-beach-honolulu-454222.html

Parsons, C. D. F. (Ed.). (1995). *Healing practices in the South Pacific.* Institute for Polynesian Studies.

Pelling, M. (2011). *Adaptation to climate change: From resilience to transformation.* Routledge.

POAWW. (2020). Protect Our Ala Wai Watersheds website. https://www.poaww.org/

Popke, J. (2016). Researching the hybrid geographies of climate change: Reflections from the field. *Area, 48*(1), 2–6.

Rose, G. (2000). Hybridity. In R. Johnston, D. Gregory, G. Pratt, & M. Watts (Eds.), *The dictionary of human geography* (pp. 364–365). Blackwell.

Rosenzweig, C., Solecki, W. D., Romero-Lankao P., Mehrota, S., Dhakal, S., & Ibrahim, S. A. (2018). *Climate change and cities: Second assessment report of the Urban Climate Change Research Network*. Cambridge University Press.

Shock, E. V. (1985). *Ho'oponopono: Contemporary uses of a Hawaiian problem-solving process*. University of Hawai'i Press.

Thieme, T. A. (2016). A reflection on resilient urban economies. The British Academy. https://www.thebritishacademy.ac.uk/documents/1881/Reflection-Resilient-Urban-Economies-Thieme.pdf

United Nations. (1992). United Nations Framework Convention on Climate Change. https://unfccc.int/resource/docs/convkp/conveng.pdf

USACE. (2020). Ala Wai Project facts. US Army Corps of Engineers (USACE) website. http://www.poh.usace.army.mil/Missions/Civil-Works/Civil-Works-Projects/Ala-WaiFlood-Risk-Management-Project/Ala-Wai-Project-Facts/

Watts, M. J. (1983). *On the poverty theory: Natural hazards research in context*. Allen and Unwin.

Webber, S. (2016). Climate change adaptation as a growing development priority: Towards critical adaptation scholarship. *Geography Compass, 10*(10), 401–413. https://doi.org/10.1111/gec3.12278

Wehe Wehe. (2019). Hawaiian electronic dictionary. https://wehewehe.org

Westley, F., Olsson, P., Folke, C., Homer-Dixon, T., & Vredenburg, H. (2011). Tipping toward sustainability: Emerging pathways of transformation. *Ambio, 40*(7), 762–780. https://doi.org/10.1007/s13280-011-0186-9

Wickson, F., Carew, A. L., & Russell, A. W. (2006). Transdisciplinary research: Characteristics, quandaries and quality. *Futures, 38*(9), 1046–1059. https://doi.org/10.1016/j.futures.2006.02.011

Williams, R. (1979). *Politics and letters: Interviews with* New Left Review. Verso.

Wright, E. O. (2010). *Envisioning real utopias*. Verso.

About the author

Valentine Huet is a transformative sustainability adviser at OpenCommunities Consulting in Paris. She accompanies private and public actors to transform their practices towards a common, desirable and suitable future using collective intelligence and systemic approaches. She completed her MSc thesis in the KTH Environmental Humanities Laboratory (EHL), based on a case study in Hawai'i that investigated the dichotomy between grassroots movement interests and government strategies.

6. Turning urban fragilities into resources for a just climate governance

Gilda Berruti and Maria Federica Palestino[1]

Abstract

Employing Naples (Italy) as their case study, Berruti and Palestino challenge the neoliberal framework of climate urbanism by adopting a "Southern" perspective. Their research shows the potential of social practices as the base for experimental governance approaches. Berruti and Palestino delves into a working-class neighborhood through in-depth interviews and a participatory laboratory, exploring the perceptions of climate change and the concrete practices of adaptation and mitigation. A systematic review of local newspapers for a three-year period (2018–2020) provides clues on the mainstream perception of climate change. Finally, the authors analyze the policies and official documents of the local institutions tackling climate change.

Keywords: urban fragility, DIY practices, perceptions, Naples

> *Cultivating a stretch of land is only a means to an end. Integration is the bulwark of our social garden.*
>
> —A. A., Lilliput Day-Care Center

> *While working in the garden we experience climate change on our skin.... In the morning we wake up and look at the sky to know if we have to water or not, and we end up becoming one with the climate.*
>
> —R. D., Era Social Cooperative

1 This contribution is the result of a joint effort by the authors. Sections 3, 5 and 6 were written by Palestino, Sections 4 and 7 were written by Berruti, and Sections 1, 2 and 8 were written by both authors.

Armiero, M., E. Turhan & S. Paolo De Rosa (eds), *Urban Movements and Climate Change: Loss, Damage and Radical Adaptation*. Taylor & Francis Group 2023
DOI: 10.5117/9789463726665_CH06

Climate urbanism is based on the assumption that cities are primary engines of economic growth and strategic sites of climate mitigation and adaptation (Long & Rice, 2018). By responding to climate change in a problem-solving framework, climate urbanism risks disregarding spatial rights and environmental justice. Such an urbanism, addressing the wealthiest regions on a global scale, is liable to build new gated communities, producing spatial injustices and green gentrification (Anguelovski et al., 2019). Due to the socio-spatial selectivity of new urban environmental regimes (Rosol et al., 2017), investments in green urbanism are city-centric and end up overlooking peripheral districts and uneven constellations of "extended urbanization" (Brenner & Schmid, 2015), not only in the Global North, but also in the Global South.

Prosperous cities, most of whose planning models were produced under the auspices of neoliberal governments, make use of powerful entrepreneurial climate coalitions and are equipped with political and technical elites that compete for and are awarded funds to conduct environmental projects. On the other hand, cities in economically depressed regions are unlikely to turn into centers of financial power or hubs for green coalitions, so they inevitably remain outside of climate urbanism developments. This is also due to local institutions' lack of technical capabilities and scarce opportunities to improve their planning skills and environmental knowledge (Berruti & Palestino, 2021).

Research on the factors that contribute to the diversity of climate change responses in European cities shows that large and prosperous cities do more climate planning than vulnerable cities (Reckien et al., 2015). This finding is confirmed by the recent report on the state of local climate planning in Europe, which indicates that "there is wide variation among countries in the prevalence of local climate plans, with generally more plans developed by central and northern European cities" (Reckien et al., 2018, p. 208).

Looking at climate urbanism from a neoliberal perspective, homogeneous trends of urban development emerge that dim plural experiments and social responses to climate change (Robin & Castán Broto, 2021). Consistent with this position, a new gaze is needed that strengthens ongoing experiments, turning climate change into an opportunity for coproducing transformation. As the case of Naples will show, some cities enjoy a different potential, one that is related to the wealth of urban social capital and the creativity of environmental grassroots initiatives.[2] A variety of nongovernmental actors,

2 The prime example of this kind of climate-responsive city in Italy is Bologna. As a matter of fact, once local institutions (from the region to the municipalities and the metropolitan city)

such as citizen collectives and movements, firms, and universities, behave as new agents of change in the city's climate governance, thus producing diverse consequences for agency and empowerment (Van der Heijden et al., 2019). These initiatives can move into transformation only when they produce civil society-based innovations, expanding even beyond the niche of social innovations they contribute to triggering (Seyfang & Haxeltine, 2012).

The social potential that challenges the neoliberal framework can be appreciated only by adopting a "Southern" perspective on climate urbanism, one that is open to seeing differences between places (Robin & Castán Broto, 2021), thus taking climate justice demands into account. Starting from Robinson's (2016) understanding of the value of "thinking cities through elsewhere," the reference to a Southern perspective calls for opening urban studies to more global insights. According to Cremaschi and Lieto (2020, p. 263), adopting a Southern perspective from a position in the Global North demands "a theoretical repositioning dealing with a plurality of different Souths, the old ones, which have never been entirely understood within the 'modern' white, protestant, imperialist condition; and the new ones suspended between colonialism and dependent developmentalism." Within a Southern European framework, climate urbanism "must attend to below-the-radar forms of action and explore their transformative potential and capacity to improve living conditions for the urban majority under climate change" (Robin & Castán Broto, 2021, p. 873).

If climate change is not reduced to problem-solving or a market strategy, but seen as a critical and severe environmental and societal problem (Stang & Ujvari, 2015; Laakso et al., 2017), experimental approaches to climate governance will be required "as a response to the perceived failure of 'command and control' regulation in a rapidly changing world,... where front line actors need to find joint solutions to common problems through experimental trial and error processes" (Morgan, 2018, p. 8).

In impoverished regions, policymakers often experience difficulties with implementing climate plans, while citizens who creatively manage environmental practices from below are unprepared to respond to the effects of climate change through civic initiatives. In the direction of an experimental governance process, they both need to learn from their mutual interactions in order to compare alternative approaches to common general aims (Sabel, 2005).

started to implement an early sustainable energy and climate action plan, a strong synergy with social actors and citizens was built, including the Extinction Rebellion movement for its climate public assemblies. See https://www.xrbologna.it.

The following sections, after introducing the urban fragility of Naples and focusing on methodology, offer a portrait of the urban region of Naples. Subsequently, a climate-responsive approach for recovering a peripheral neighborhood in eastern Naples is explored, with a deep inquiry into the Social Garden of Health experiment. This inquiry shows the potential to use social practices as the base for experimental governance approaches. Our conclusions generalize the possibility of channeling the transformative potential of climate change in order to update climate governance by means of experimentation, supporting different skills and needs and enhancing rights.

Urban fragility as a lever

As engaged scholars at the University of Naples Federico II, in Italy, we have explored the working of the public sector in the metropolitan area of Naples, and how different stakeholders are reacting to the inefficacy of public policies addressing climate change in the context of the current economic depression. From this situated perspective, we argue that cities that are cut off from global projects and funds gain the potential freedom to seize climate change as an opportunity to make public policies responsive to the redistribution of environmental goods and rights. This can be made possible by facilitating encounters between expert and contextual knowledge—in other words, by enhancing technical knowledge, public/private know-how and capabilities, and civic education through participation and interaction among stakeholders.

According to Muggah (2016), urban fragility occurs "when city authorities are unable or unwilling to deliver basic services to citizens" and "is triggered by a rupture of a city's social contract." We see Naples as a perfect case of such urban fragility where loss and damage from climate change are the basis for recovering places and communities from below. Our claim is that urban fragilities can be turned into resources for innovation in the climate governance of economically depressed cities and progress toward community-based adaptation pathways.

In order to corroborate this hypothesis, we adopt the concept of "transformation as liberation" as a frame of reference that challenges the status quo and addresses the root causes of vulnerability (Roberts & Pelling, 2020). This change requires decision-makers to reframe the governance model in use, promoting the social turn of climate policies through public investments in experimental climate governance (Sapiains et al., 2021). This experimental governance is based on the public sector's ability to learn by monitoring the process and on local actors' broad discretion to pursue their objectives (Sabel & Zeitlin, 2012). Furthermore, as Castán Broto and Bulkeley (2013, p. 101) emphasize, "experimentation involves multiple forms of technical and social innovation."

Methodology

We adopted a case-study methodology to emphasize the degradation and pollution of Ponticelli, a former industrial neighborhood of Naples where the inhabitants have been waiting for reclamation since the 1990s. Located on the eastern plain far from the core of Naples, Ponticelli remains unknown to the majority of the city's citizens, despite its celebrated rural past.[3]

A qualitative inquiry and mapping (Lynch, 1960) aimed at reclaiming the urgent recovery of the eastern plain was integrated into a policy analysis approach (Dente, 2013) in order to explore the local policymaking and the governance model. First, according to Kevin Lynch's "image-survey" methodology and other insights in *A Theory of Good City Form* (1981), a qualitative inquiry conducted in-depth interviews with 105 inhabitants. The interviews focused on the relationship between inhabitants and the progressive deterioration of the humid site. Thirteen questions featured in the interview were designed to determine the residents' opinions of the "performance dimensions" of *vitality, sense* and *fit* (Palestino, 2016). After the survey, the residents' responses were coded and represented using collaborative mapping to territorialize the people's feelings about the loss of rural culture and damage due to the proliferation of brownfields. Starting with the collection of people's memories of the preindustrial site, the preconditions for community-based regeneration pathways were explored.

Second, a participatory laboratory was carried out to determine local perceptions of climate change impacts. It was intended to coproduce knowledge through close interactions between skilled researchers in the field of climate change adaptation and mitigation and inhabitants with skills and experience derived from their places of residence. Third, a textual

3 Although the fluvial landscape of the marshy plain is one of the dominant themes in Neapolitan historical iconography, it is noteworthy that the contribution of teachers, professionals and amateur scholars with a passion for rural material culture and tradition on the east side of the city has also fueled an interesting "pop" trend in landscape studies, allowing the memory of the humid zone and the vegetable gardens to remain alive in the local community.

and content analysis of articles in local newspapers[4] for a three-year period (2018–2020) was done in order to understand how the global challenge of climate change was being described by the local media, perceived by ordinary citizens and decision-makers and put on the political agenda (Palestino et al., 2020). Fourth, an analysis of the main formal decisions regarding climate change was conducted, along with interpretation of relevant planning documents. In addition, we carried out about 20 in-depth interviews or group interviews about climate perception with key informants, chosen from local politicians and officials, professionals, members of associations and social movements, and green entrepreneurs. We also collected 275 questionnaires to see how the effects of climate change were perceived by students at local schools and universities. All the collected materials were useful not only for determining general knowledge about climate change in the urban region of Naples, but also to promote, in February 2021, a public consultation of key actors for the launch of the People's Climate Plan for Naples, following the example of the People's Climate Plan for New York (Dawson, 2017; Climate Action Lab, 2019).

A portrait of the urban region of Naples

Naples is the capital of a metropolitan area of 92 municipalities, with over 3 million inhabitants in 1,171 square kilometers. The city, with about 955,000 inhabitants living in an area of 119 square kilometers, is the regional capital of Campania (550 municipalities and 5.827 million inhabitants). It is located in the heart of southern Italy, also known as the Mezzogiorno. To specify the current condition of Naples as part of the economy and society of southern Italy, we refer to the report by Svimez, the Association for the Industrial Development of Southern Italy (Svimez, 2019), which identified a double gap in drawing its socioeconomic profile. The gap consists in the socioeconomic distance between north and south of Italy, on the one hand, and between Italy and the rest of Europe, on the other. This distance increased with the pandemic, emphasizing the unsolved structural knots that united Italian regions during the COVID-19 crisis and separated during the recovery phase (Svimez, 2021).[5]

4 The newspapers were *La Repubblica-Napoli* and *Il Mattino-Napoli*.

5 As reported by Svimez (2021), it is time to tackle the structural deficiencies that have been hindering national growth for 20 years and exacerbating the territorial inequalities that were already increasing before the pandemic.

In the last 20 years, Italy's national economic policy has reduced investment in the south of Italy and eroded the interdependencies of the south with the center-north, thus weakening Italian domestic markets. Interdependency among regions, together with its implications for national territorial cohesion, seem to be key aspects to focus on in order to enhance the development of southern regions. Here, in fact, the deficiencies in the infrastructure provision and service allocation are massive. Between 2006 and 2017, the GDP growth in Europe was associated with a decrease in Italian regions. Looking deeper, southern Italian cities are poorer than center-north Italian cities.

This phenomenon highlights a complex geography, characterized by development at different speeds for Central and Northern Europe, the new Eastern European states and weak regions in Mediterranean Europe and illustrating how distant Southern Europe remains from European standards of living. The COVID-19 emergency further exacerbated these inequalities, since the drop in GDP in the region of Campania was slightly higher than that registered by other southern Italian regions, as were its decreases in consumption and private investments. Further fragilities emerge if we observe the regional data on the employment rate (less than 60%). The younger generations' expectations of the future are not reassuring, since the percentage of people in the age group of 18 to 24 years who neither work nor study exceeds 22.5% in Campania and Mediterranean Europe, while the rate of employment for young graduates is less than 60% (Svimez, 2019). In both cases, these values diverge radically from those of the rest of Europe. This is also true for women's employment rate because Campania and southern Italy in general have the highest gender employment gaps in Europe.

This fragile context mirrors the spatial and social features of the Neapolitan metropolitan area. Naples is located in a densely populated and disorderly urbanized area characterized by high land take that reaches as high as 34% in the metropolitan area and 63% in the capital.[6] A low public health index and poor habitat quality (SNPA, 2021) testify to increases in human and nonhuman sickness (Armiero & Fava, 2016) and in land degradation (Corona, 2015). Being strongly affected by the recent regional waste emergency (Pasotti, 2010; Cantoni, 2016) and, as a consequence, by the ongoing socio-ecological disaster known as the Land of Fires (Armiero, 2014, 2021; Palestino, 2015), the area is now characterized by a high percentage

6 The land take indicator refers to the area of agricultural, forest and other semi-natural land taken for urban and other artificial land development (see EEA, 2019).

and variety of wastelands, and the proliferation of wastescapes waiting for regeneration (Amenta & Van Timmeren, 2018).

Administrative unregulation (De Leo & Palestino, 2017) and the spread of urban informalities (Berruti & Palestino, 2020) are key aspects to understand the context, with impacts on the weak governance model in use, that is marked by fragmented decision-making processes and sectoral knowledge, and failure to integrate stakeholders in policymaking. The current efforts to manage the ongoing transition are uncovering structural problems of the administrative apparatus that need to be rectified, such as financial instability, reduced staff numbers and limits to horizontal and vertical governance related to difficulties with inter-institutional communication (Berruti & Palestino, 2021).

According to a legal measure adopted to mitigate financial instability, the city of Naples has been involved since 2013 in a 10-year financial recovery plan, whose impacts are harshly affecting the staffing plan and, as a consequence, are weakening the performance of the governance model in use, and even the functioning of the administrative apparatus. The need for containing the public expenditure caused, in fact, a prompt cutting of jobs. The number of public employees fell from 13,901 to 10,474 in 2013. It decreased to 6,452 in January 2019 and to 5,740 in January 2020 (Comune di Napoli, 2019b, 2020). More than 50% of the city's employees are older than 60, and more than 75% are older than 50.

One of the most obvious weaknesses of Naples' public sector is the service devoted to the maintenance of urban greenery. Many public open spaces and parks are partially closed or neglected, due to the sharp reduction in the number of gardeners "from 1,000 in 2011 to about 60 today" according to the mayor.[7] Urban decline together with the welfare state crisis, diluting the pervasiveness of the public sector and strengthening the role of civil society in the delivery of public services (Gilbert, 2002) have led to dysfunctions, lack of services and inequalities among urban districts. These inequalities are harsher in peripheral districts, where public open spaces are often neglected and climate impacts are more devastating, calling for mixed partnerships to manage green spaces.

Despite this picture of manifold weaknesses, a rich and diverse social infrastructure is emerging in specific areas of the urban region through the involvement and commitment of local communities along with the

7 These quotes from the mayor are from the transcript of a filmed interview given to *Il Mattino*, April 15, 2021 (translation by the author), https://www.ilmattino.it/video/glocal/de_magistris_parla_privatizzazione_villa_comunale_non_ha_capito_niente-5901069.html.

contribution of the University of Naples Federico II, joining the urban arena in the name of public engagement. They all provide a powerful social infrastructure that, if supported, could improve the quality of the climate governance model in use and design policies tailored to the context.

This portrait of the city explains the reason why climate change is not a priority for the public urban agenda. For fragile cities such as Naples, bringing global issues to the local level is difficult due to a "readiness argument" that stresses the perception of local deficiencies, thus hindering local administrations' ability to address global environmental concerns (Bai, 2008). Accordingly, the most urgent issues in Naples are related to the current financial and welfare crisis, and difficulties with the management of ordinary business (Berruti et al., 2022). However, the issue of climate change is gradually taking a symbolic and educational meaning, not only for citizens, but also for improving the city and the metropolitan area's technical competences, know-how and upgrading its governance model (Berruti & Palestino, 2021). Fortunately, social organizations, cooperatives and civic groups have a deep and long-lasting presence in the metropolitan area of Naples (Palestino, 2012; Berruti & Palestino, 2014). In fact, they have also often functioned as a powerful lever for stopping decline during the recent structural crisis and environmental emergencies (De Rosa, 2017).

A climate-responsive approach for a peripheral neighborhood

As urban planning scholars, working for a long time on fragile areas, we have researched how to regenerate the eastern plain of Naples, which includes the Ponticelli neighborhood, by involving the communities that live there. In so doing, climate change seemed an opportunity for us to recover abandoned or unfinished public spaces and services by means of physical adaptation and social empowerment. Convinced that "the environment of the city (both social and physical) is the result of a historical-geographical process of the urbanization of nature" (Swyngedouw & Kaika, 2000, 569), we put a specific focus on the transition of the plain from the city's most fertile agricultural land to the post-industrial settlement it is today.

During a centuries-long process of reclamation and drainage, the eastern plain, which had been a marshland, became the vegetable garden of Naples (Di Martino, 2013). Due to its distance from the city's center, Ponticelli remained a farming community for a long time, while industrialization and a rapidly increasing resident population grew next to it over the course of a few decades. In 1968, a new site for public housing was planned for

Ponticelli, and after the 1980 earthquake, it was confirmed as one of the target areas for the Emergency Housing Plan, providing housing and facilities for citizens who had lost their homes (Pagano, 2001). Cut off from the core of the city and due to poor transportation infrastructure on the slopes of Mt. Vesuvius, the area was abandoned to a one-way destiny of marginality and decline.

Since deindustrialization in the 1990s, Ponticelli became one of the city's most vulnerable ecosystems, both ecologically and socially. Here, in fact, facing environmental risks is an everyday fact of life, due to a complex interaction between infrastructural barriers, the depletion of industrial soils, the low quality of the public housing stock, highly polluted sites awaiting reclamation and deprived populations. This is why our field research was intended to test climate-responsive approaches for planning methodologies tailored to peripheral areas and communities. From the very beginning, therefore, we focused on whether (and how) climate change could offer an opportunity to couple climate studies with urban inquiries into local communities' perceptions, needs and expectations.

Although in Ponticelli, agriculture has ceased to be part of people's know-how since about the 1960s (Palestino, 2013), today the rural tradition can be advocated and recovered only in the light of contemporary needs and lifestyles. According to this thesis, we have recently selected the neighborhood as the center stage for collaborative experiments focused on climate education and community resilience. With this goal in mind, Ponticelli and the eastern plain have been described and represented as urban ecosystems that can be recovered through climate adaptation and transformation, i.e., by creating community-based regeneration pathways based on horticulture.

The Lynchian map *Living Memories of the Rural Past in Napoli* (Figure 6.1), indicates the persistent bond between Ponticelli's inhabitants and its rural past, despite the industrial shift of the 1960s and the closedown of its factories in the 1990s. The map also shows the importance of water and its infrastructure in the culture of the plain and the people's image of it. What the map represents are, accordingly, unconscious feelings of "loss and damage" (Roberts & Pelling, 2020) for habits, places and behavioral settings that have been washed away while industrializing the settlement. The current image of the neighborhood is not as detailed as its past image, which proves that interviewees do not care for the current urban environment.

As planners and designers, we felt doubly involved in advocating the popular right to urban environments (Kaika & Swyngedouw, 2011), both for supporting claims to justice and for circulating innovative methodologies to

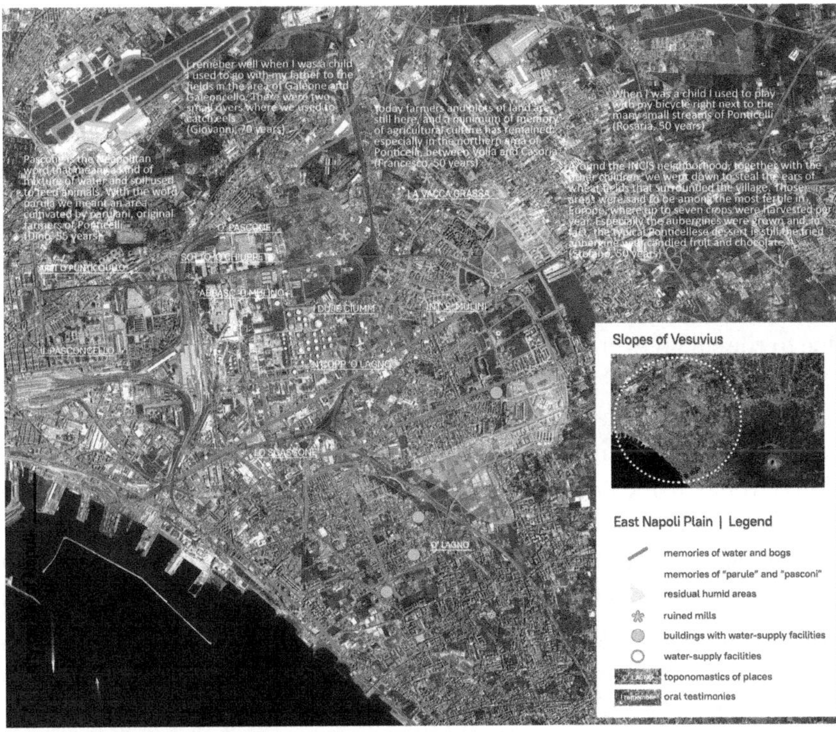

Figure 6.1. *Living Memories of the Rural Past in Napoli*, by Maria Federica Palestino. Graphic design by Marilena Prisco.

share the climate change turn in urban studies. Narratives and maps about the rural past uncover the persistence of Ponticelli's preindustrial identity and constitute a link to future actions in the name of rural culture as shown by our case study of the Social Garden of Health (Orto sociale della salute).

Combating the local effects of climate change became the pretext to experiment, by means of a participatory laboratory, with a strategy of reclaiming environmental justice and improving the neighborhood's livability and social inclusion, thus activating opportunities for "transformation as liberation" from loss and damage, and producing inclusive decision-making (Roberts & Pelling, 2020).

The Social Garden of Health as an unconscious reaction to loss and damage

During the participatory laboratory on climate change, a group of stakeholders from Ponticelli showed us the local practice of the vegetable garden located in the Fratelli de Filippo Park, a public facility partially closed between 1995 and 2008 due to a lack of public management and the city's financial breakdown. In 2015, after repeatedly failing to bring it back to public use, the city gave up, and the abandoned park was entrusted by the city council to the local health authority. The latter has been involved in the temporary management of the park by means of an agreement with the city of Naples that allows collective maintenance of its open spaces with the direct involvement of the local community.[8]

The popular rediscovery of horticulture in this public area was largely due to the social training by the health authority through Lilliput Daycare Center, which is still charged with reintegrating drug addicts into social life. The basic idea was that addiction cannot be resolved only within health facilities, but it requires reintegration into the community. Vulnerable people today work side by side with other groups in the garden, learning integration and transcending social biases by doing so.

The idea behind the social garden consisted in recreating rules and routines by means of which the community could take care of open spaces that had gradually lost their original public use, becoming relicts from a past season of the welfare state. The aim was to extend the reopening of the park to associations, schools, parishes and individual citizens, so as to nurture a caring relationship connecting rehabilitated drug users and the local community. As a consequence, inhabitants and local operators and educators were given an open invitation to adopt a parcel of the abandoned area in order to recover the park as a "Social Garden of Health."

The intellectual and material resources for the project came from an innovative approach to health policies that merges the rehabilitation of vulnerable people with spatial regeneration. The major social actors involved in the Fratelli de Filippo Park are the social cooperative Era, training drug addicts to become social gardeners, and the consortium of social cooperatives Gesco Campania, which has developed a powerful model for the participative management of the abandoned urban green spaces. The resurgence of the public park as a vegetable garden was successful thanks to the idea of taking back the land that is fully consistent with revisiting local feelings and social behaviors in a contemporary key. The people's rediscovery of land consisted, indeed, in enhancing local culture by compensating for

8 The temporary management system is among the experimental tools with which the local administration of Naples has planned to comanage the city's common goods with civil society. See: Comune di Napoli, 2019a; Capone, 2021.

Figure 6.2. Garden in Ponticelli. Photo by Gianni Fiorito.

injustices due to losses of rural identity, and damages due to the urban impacts of industrialization.[9]

Collective gardening not only rehabilitated drug addicts, but even recovered the largest public space inside the neighborhood from its abandonment by the public administration. While regaining their gardening culture (Figure 6.2), the inhabitants were involved in a participatory laboratory on the local perception of climate change that revealed an opportunity for co-experimenting with how to turn the popular consciousness of gardening into practices of "do-it-yourself (DIY) adaptation" (Cloutier et al., 2018). During the laboratory, we helped to sustain the popular resurgence of rural culture by means of interactive tools such as collaborative mapping, video storytelling, codesigning and building adaptation devices,[10] turning participants into key players in the narration of Ponticelli's socio-ecological transition to the rest of the city (Palestino, 2017; Visconti, 2017).

9 Between 2006 and 2012, by effect of Act 3/4/2006 n. 152, 57 Sites of National Interest (Siti di Interesse Nazionale) were selected for reclamation by the Ministry of the Environment. One of them is located on the eastern plain of Naples near the Ponticelli neighborhood. Due to the length and complexity of the administrative procedures involved, very little has been done to reclaim these areas.

10 To get an idea of the interactive tools and practices used in the participatory laboratory, watch the documentary *Resilient Ponticelli: Tackling Climate Change with the Community* (Palestino & Visconti, 2017).

We were confident that, thanks to this experimental communicative framework, institutions could observe and listen to local voices, considering Ponticelli as an appropriate context where different experiments could be compared and interact, monitoring and connecting each other. In our opinion, in fact, codesigning a participatory process with local groups and adopting an innovative plan of communication might have pushed the neighborhood to overcome the concatenation of malignant effects that had produced "loss and damage" and allow for the reappropriation of public spaces and the transformation of the inhabitants' living conditions. The experience of the laboratory can be framed, accordingly, as an attempt at experimental governance, working on how to move from "do-it-yourself adaptation" to "community-based adaptation" (Dodman & Mitlin, 2013) as an opportunity for transforming insurgent environmental practices into collaborative climate policies. Although the Social Garden of Health experiment became well known in metropolitan Naples, the learning-by-monitoring approach by the public sector still needs improvement, while local actors' discretion in creatively managing public open spaces calls for support by institutions in order to improve the climate agenda.

Moving from social practices to experimental governance

In 2018, the lack of perceptions of climate impacts on the urban environment of Naples seemed to be largely due to mistakes in global media communication and governmental approaches that resulted in pushing climate change out of public discourse. As a first step, we explored how climate change was perceived in the urban arena: from public institutions to politicians, expanding to associations, movements, activists, social operators and cooperatives, and then even to students at schools and universities. We discovered that climate change was not perceived as a priority, but usually seen as referring to exotic narratives such as a polar bear traveling on the tip of an iceberg or attributed to technocratic responses in the jargon of "tons of CO_2 emissions." More alarming emergencies, due to the spread of poverty and high rates of unemployment, and only a few climatic effects on the city and its communities, strengthened this common perception (Palestino et al., 2020).

More generally, however, the proliferation of different public discourses has the effect of confusing imaginaries, making climate change function as a conundrum (Swyngedouw, 2010). On the one hand, global measures and strategies succeed in politicizing climate change, although keeping it at a distance from ordinary people. On the other, prevailing feelings of being defenseless in the face of something out of individual control and independent even from collective behaviors have the effect of depoliticizing climate change.

In Naples, climate change narratives initially had the effect of maintaining the climate at a distance from both citizens and public authorities so as to completely deterritorialize the issue. This situation began to change in February 2019,[11] when climate change suddenly grabbed the media's attention due to Greta Thunberg's commitment as the inspiration for the Fridays for Future movement, and the topic unexpectedly became part of the Neapolitan political agenda, too.

Analysis of the climate governance in the urban region of Naples shows that the first global march for climate and the resulting media attention worked to activate both urban and metropolitan policies, thanks to the collaboration between the University of Naples Federico II and local administrations (Berruti & Palestino, 2021).[12] This collaboration, integrating environmental concerns in city and metropolitan planning, supported the elaboration of the Preliminary Urban Plan of Naples, and the activation of the Metropolitan Strategic Plan, thus dealing with the city and metropolitan area's unpreparedness to respond to climate impacts.

A mutual learning process was initiated, involving the public sector and the University of Naples Federico II, which created an open space where other actors could also participate. This learning process slowly proceeded, on the one hand, by developing technical and political capabilities to respond to climate change in the public sector, thus dealing with knowledge gaps and administrative inefficacy, and, on the other, by sharing climate responsive narratives and actions in the metropolitan area, with actors enhancing each other's commitment.

In order to move from a single experiment to an experimental governance approach, institutions have to play a key role, working as guarantors, in the interplay among varied partners from the public, private and civic sectors. In addition, starting with each experiment, strict coordination between the city and the metropolitan and regional governments is needed. In this framework, the role of researchers and scholars as promoters of environmental values

11 In particular, the date of February 15, 2019, can be seen as a divide, although more than 20 years of grassroots activism, Global South climate politics, climate camps and scientists' warnings had already transpired.

12 Interviews with city and metropolitan area officials provide evidence of the boost given by the media attention to climate change.

and socio-technical innovation cannot be overlooked. Social gardening and ethical farming practices in the region, which occur without being aware of their status as potential responses to climate change, have to be detected and inventoried as a first step toward setting up experimental governance approaches. Then, the experiments can become cross-sectoral and involve the entire city as we tested in launching the People's Climate Plan for Naples, promoted by the Department of Architecture in February 2021, involving an open discussion among creatives, institutions, public planning, environment and health officials, academics, high-school teachers, nongovernmental organizations, cooperatives, movements and opinion leaders (Armiero et al., 2021). This event was only the first of a series, based on the idea that sharing different experiences, competences and feelings concerning climate change could be transformative by producing sociocultural innovations.

Conclusions

Experimental governance approaches to climate change can be useful in the case of fragile cities, where the public sector has difficulties with ensuring the implementation of climate policies, while active groups of citizens, involved in innovative environmental experiments, are not ready to respond to climate change impacts. The city of Naples has the potential to respond to climate change effects on the urban fabric and human and nonhuman communities. This potential emerges from a Southern perspective when a lack of knowledge, technical abilities and public funds are replaced with the richness of the social infrastructure. Starting with this out-of-necessity resource, transformation can be promoted and advanced based on development pathways that address the root causes of vulnerability, investing in site specificity and considering social justice demands. A plurality of actors from public, private and social arenas is involved in this transformation, adopting an experimental governance approach, where a trial-and-error methodology drives ways of governing the city.

One of the greatest challenges in economically depressed cities, which cannot afford sophisticated equipment, is overcoming the tension between austerity and sustainability—that is to say, managing the public sector's weaknesses due to budget cuts, on the one side, and, on the other, the demand for a public actor that can act as a guarantor and bring innovation to ordinary routines. In this framework, climate change becomes a lens through which to read different actors' potentials for innovation, mainly in public spaces, and a trigger for new challenges and mutual comparison.

Adopting a Southern European perspective on climate helps different sensitivities to emerge and injustices to be faced. This can occur through comanaged experiments that can be improved and networked using incremental adjustments. Experiments with climate change responses, if carefully assessed and networked, can produce long-term changes and transformations, thanks to the involvement of actors and groups and their commitment to address local vulnerabilities and injustices.

References

Amenta, L., & Van Timmeren, A. (2018). Beyond wastescapes: Towards circular landscapes—Addressing the spatial dimension of circularity through the regeneration of wastescapes. *Sustainability, 10*(12), 4740. https://doi.org/10.3390/su10124740

Anguelovski, I., Connolly, J. J., Garcia-Lamarca, M., Cole, H., & Pearsall, H. (2019). New scholarly pathways on green gentrification: What does the urban "green turn" mean and where is it going? *Progress in Human Geography, 43*(6), 1064–1086. https://doi.org/10.1177/0309132518803

Armiero, M. (2014). Garbage under the volcano: The waste crisis in Campania and the struggles for environmental justice. In M. Armiero & L. Sedrez (Eds.), *A history of environmentalism* (pp. 167–184). Bloomsbury.

Armiero, M. (2021). *Wasteocene: Stories from the global dump.* Cambridge University Press.

Armiero, M., & Fava, A. (2016). Of humans, sheep, and dioxin: A history of contamination and transformation in Acerra, Italy. *Capitalism Nature Socialism, 27*(2), 67–82. https://doi.org/10.1080/10455752.2016.1172812

Armiero, M., Berruti, G., Madonna, R., Palestino M. F., Anguelovski, I., Dawson A., Sedrez, L., & Spaccaforno P. (2021). Occupy Climate Change, città dopo città. In Occupy Climate Change (Ed.), *Trame Pratiche e saperi per un'ecologia politica situata Occupy Climate Change* (pp. 91–112). Tamu edizioni.

Bai, X. (2008). Integrating global environmental concerns into urban management: The scale and readiness argument. *Journal of Industrial Ecology, 11*(2), 15–29. https://doi.org/10.1162/jie.2007.1202

Berruti, G., & Palestino, M. F. (2014). The park of Scampia: Notes on a co-management experiment. *Urbanistica Informazioni*, 257, 136–139.

Berruti, G., & Palestino, M. F. (2020). Contested land and blurred rights in the Land of Fires (Italy). *International Planning Studies, 25*(3), 277–288. https://doi.org/10.1080/13563475.2019.1584551

Berruti, G., & Palestino, M. F. (2021). Exploring the governance of Naples, Italy, through a climate responsive approach. In E. Peker & A. Ataov (Eds.), *Governance of climate responsive cities: Exploring cross-scale dynamics* (pp. 43–58). Springer Nature.

Berruti, G., Laino, G., Mattiucci, C., & Palestino, M. F. (2022). Napoli: i beni comuni come dispositivo per accompagnare le transizioni. In A. Magnier, M. Morisi, & C. Perrone (Eds.), *Urban@it Settimo Rapporto sulle città Chi possiede la città? Proprietà, poteri, politiche* (pp. 167–183). Il Mulino.

Brenner, N., & Schmid, C. (2015). Towards a new epistemology of the urban? *City, 19*(2–3), 151–182. https://doi.org/10.1080/13604813.2015.1014712

Cantoni, R. (2016). The waste crisis in Campania, South Italy: A historical perspective on an epidemiological controversy. *Endeavour, 40*(2), 102–113. https://doi.org/10.1016/j.endeavour.2016.03.003

Capone, N. (2021). Dispositivi giuridici per la città pubblica e l'uso in comune dello spazio urbano. L'esperienza napoletana dei beni comuni. In C. Perrone, B. Masiani, & F. Tosi (Eds.), *Una geografia delle politiche urbane tra possesso e governo. Sfide e opportunità nella transizione* (pp. 212–223), Working Papers-Urban@it.

Castán Broto, V., & Bulkeley, H. (2013). A survey of urban climate change experiments in 100 cities. *Global Environmental Change, 23*(1), 92–102. https://doi.org/10.1016/j.gloenvcha.2012.07.005

Climate Action Lab. (2019). *A people's climate plan for New York City?* The Graduate Center, City University of New York. https://occupyclimatechange.net/wp-content/uploads/2022/03/ClimatePlan_FINAL1-1.pdf

Cloutier, G., Papin, M., & Bizier, C. (2018). Do-it-yourself (DIY) adaptation: Civic initiatives as drivers to address climate change at the urban scale. *Cities, 74*, 284–291. https://doi.org/10.1016/j.cities.2017.12.018

Comune di Napoli. (2019a). Napoli 2019-2030 Città ambiente, diritti e beni comuni Preliminare del Piano Urbanistico Comunale. https://www.comune.napoli.it/flex/cm/pages/ServeAttachment.php/L/IT/D/1 percent252F7 percent252Ff percent252FD.e499f9cd25665caa4e98/P/BLOB percent3AID percent3D37912/E/pdf-

Comune di Napoli. (2019b). Documento Unico di programmazione 2019-2021. Sezione operativa. https://www.comune.napoli.it/flex/cm/pages/ServeBLOB.php/L/IT/IDPagina/39974-

Comune di Napoli. (2020). Documento Unico di programmazione 2019-2021. Sezione operativa. https://www.comune.napoli.it/flex/cm/pages/ServeBLOB.php/L/IT/IDPagina/41877

Corona, G. (2015). Questione meridionale come questione territoriale. Il caso della Campania. *Parolechiave, 54*, 153–165.

Cremaschi M., & Lieto L. (2020). "Writing Southern theory from the Global North. Notes on informality and regulation", *Equilibri*, Special Issue, 24, 261-80, doi: 10.1406/98117

Dawson, A. (2017). *Extreme cities: The peril and promise of urban life in the age of climate change*. Verso.

Dente, B. (2013). *Understanding policy decisions*, Springer.

De Rosa, S. P. (2017). *Reclaiming territory from below: Grassroots environmentalism and waste conflicts in Campania, Italy*. Phd diss., Lund University.

De Leo D., & Palestino M.F. (2017). S-regulation matters. In A. Balducci, V. Fedeli, & F. Curci (Eds.), *Post-Metropolitan Territories. Looking for a New Urbanity* (pp. 274-280). Routledge, London-New York.

Di Martino, D. (2013). Bonifica e riciclo della piana campana. *Piano Progetto Città*, 27–28, 212–231.

Dodman, D., & Mitlin, D. (2013). Challenges for community-based adaptation: Discovering the potential for transformation. *Journal of International Development, 25*(5), 640–659. https://doi.org/10.1002/jid.1772

EEA. (2019). Land take in Europe. European Environmental Agency. https://www.eea.europa.eu/data-and-maps/indicators/land-take-3

Gilbert, N. (2002). *Transformation of the welfare state: The silent surrender of public responsibility*. Oxford University Press.

Kaika, M., & Swyngedouw, E. (2011). The urbanization of nature: Great promises, impasse and new beginnings. In G. Bridge & S. Watson (Eds.), *The new Blackwell companion to the city* (pp. 96–107). Wiley-Blackwell.

Laakso, S., Berg, A., & Annala, M. (2017). Dynamics of experimental governance. A meta-study of functions and uses of climate governance experiments. *Journal of Cleaner Production, 169*, 8–16. https://doi.org/10.1016/j.jclepro.2017.04.140

Long, J., & Rice, J. L. (2019). From sustainable urbanism to climate urbanism. *Urban Studies*, 56(5), 992–1008. https://doi.org/10.1177/0042098018770846

Lynch, K. (1960). *The image of the city*. MIT Press.

Lynch, K. (1981). *A theory of good city form*. MIT Press.

Morgan, K. (2018, December 14). Experimental governance and territorial development. Background paper for OECD/EC Workshop "Broadening innovation policy: New insights for regions and cities," Paris. https://www.oecd.org/cfe/regionaldevelopment/Morgan(2018)ExperimentalGovernanceAndTerritorial-Development_OECD_FINAL.pdf

Muggah, R. (2016, February 9). How fragile are our cities? World Economic Forum. https://www.weforum.org/agenda/2016/02/how-fragile-are-our-cities

Pagano, L. (2001). *Periferie di Napoli. La geografia, il quartiere, l'edilizia pubblica*. Electa.

Palestino, M. F. (2012). *Immaginazioni. Materiali per costruire strategie promozionali inclusive*. CLEAN.

Palestino, M. F. (2013). A survey on community resilience. In F. D. Moccia & M. F. Palestino (Eds.), *Planning stormwater resilient urban open spaces* (pp. 56–64). CLEAN.

Palestino, M. F. (2015, July 13–16). How to put environmental injustice on the planner's radical agenda: Learning on the Land of Fires—Italy. In M. Macoun & K. Maier (Eds.), *Definite space–fuzzy responsibility: Book of proceedings of the 29th AESOP 2015 Congress* (pp. 2576–2586). Prague. https://cupdf.com/document/aesop-2015-book-of-proceedings.html

Palestino, M. F. (2016). The collaborative mapping of eastern Naples: Between climate change and community resilience. In V. D'Ambrosio & M. F. Leone (Eds.), *Progettazione ambientale per l'adattamento al Climate Change. 1 Modelli innovativi per la produzione di conoscenza/Environmental design for climate change adaptation, 1: Innovative models for the production of knowledge* (pp. 158–168). CLEAN.

Palestino, M. F. (2017). Ponticelli smart lab: A hybrid environment for the implementation of experimental approaches to climate change. In V. D'Ambrosio & M. F. Leone (Eds.), *Progettazione ambientale per l'adattamento al Climate Change 2/Environmental design for climate change adaptation, 2* (pp. 132–138). CLEAN.

Palestino, M. F., & Visconti, C. (2017). *Resilient Ponticelli: Tackling climate change with community* [video]. https://vimeo.com/246951600

Palestino, M. F., Berruti, G., & Quagliano, S. (2020). Climate change as a lever for place-based regeneration policies: The case of Naples, Italy. In *Proceedings of the conference "Production of Climate Responsive Urban Built Environments," May 22–24, 2019, Istanbul* (pp. 83–89). Istanbul Policy Center.

Pasotti, E. (2010). Sorting through the trash: The waste management crisis in southern Italy. *South European Society and Politics, 15*(2), 289–307. https://doi.org/10.1080/13608740903497733

Reckien, D., Flacke J., Olazabal M., & Heidrich, O. (2015). The influence of drivers and barriers on urban adaptation and mitigation plans: An empirical analysis of European cities. *PLoS ONE, 10*(8), e0135597. https://doi.org/10.1371/journal.pone.0135597

Reckien, D., Salvia, M., Heidrich, O., Church, J. M., Pietrapertosa, F., De Gregorio-Hurtado, S., D'Alonzo, V., Foley, A., Simoes, S. G., Lorencová, E. K., Orru, H., Orru, K., Wejs, A., Flacke, J., Olazabal, M., Geneletti, D., Feliu, E., Vasilie, S., Nador, C., & Dawson, R. (2018). How are cities planning to respond to climate change? Assessment of local climate plans from 885 cities in the EU-28. *Journal of Cleaner Production, 191*(1). 207–219. https://doi.org/10.1016/j.jclepro.2018.03.220

Roberts, E., & Pelling, M. (2020). Loss and damage: An opportunity for transformation? *Climate Policy, 20*(6), 758–771. https://doi.org/10.1080/14693062.2019.1680336

Robin, E., & Castán Broto, V. (2021). Towards a postcolonial perspective on climate urbanism. *International Journal of Urban and Regional Research, 45(5), 869–878.* https://doi.org/10.1111/1468-2427.12981

Robinson, J. (2016). Thinking cities through elsewhere: Comparative tactics for a more global urban studies. *Progress in Human Geography, 40*(1), 3–29. https://doi.org/10.1177/0309132515598025

Rosol, M., Béal, V., & Mossner, S. (2017). Greenest cities? The (post-)politics of new urban environmental regimes. *Environment and Planning, 49*(8), 1710–1718. https://doi.org/10.1177/0308518X17714843

Sabel, C. (2005). A real time revolution in routines. In C. Heckscher & P. S. Adler (Eds.), *The firm as a collaborative community* (pp. 106–156). Oxford University Press.

Sabel, C., & Zeitlin, J. (2012). Experimentalist governance. In D. Levi-Faur (Ed.), *The Oxford handbook of governance* (pp. 169–183). Oxford University Press.

Sapiains, R., Ibarra, C., Jiménez, G., Blanco, G., Moraga, P., & Rojas M. (2021). Exploring the contours of climate governance: An interdisciplinary systematic literature review from a Southern perspective, *Environmental Policy and Governance, 1*, 46–59. https://doi.org/10.1002/eet.1912

Seyfang, G., & Haxeltine, A. (2012). Growing grassroots innovations: Exploring the role of community-based initiatives in governing sustainable energy transitions. *Environment and Planning C: Government and Policy, 30*(3), 381–400. https://doi.org/10.1068/c10222

SNPA. (2021). Sistema nazionale per la protezione dell'ambiente—Consumo di suolo, dinamiche territoriali e servizi ecosistemici, Report di sistema, 22. https://www.snpambiente.it/wp-content/uploads/2021/07/Rapporto_consumo_di_suolo_2021-1.pdf

Stang, G., & Ujvari, B. (2015). Climate change as a "wicked problem." *European Union Institute for Security Studies, 52*, 1–2. https://www.iss.europa.eu/sites/default/files/EUISSFiles/Alert_52_Climate_change.pdf

Svimez (2019). "Il Mezzogiorno nella nuova geografia europea delle disuguaglianze", *Note di sintesi*, online at: http://lnx.svimez.info/svimez/wp-content/uploads/2019/11/rapporto_svimez_2019_sintesi.pdf (November 2022).

Svimez (2021). *Sintesi del Rapporto SVIMEZ 2021*, online at: http://lnx.svimez.info/svimez/sintesi-del-rap¬porto-svimez-2021/ (November 2022).

Swyngedouw, E. (2010). Apocalypse forever? Post-political populism and the spectre of climate change. *Theory, Culture & Society, 27*(2–3), 213–232. https://doi.org/10.1177/0263276409358728

Swyngedouw, E., & Kaika, M. (2000). The environment of the city or … the urbanisation of nature. In G. Bridge & S. Watson (Eds.), *A companion to the city* (pp. 567–580). Wiley-Blackwell.

Van der Heijden, J., Certomà, C., & Bulkeley, H. (2019). Promises and concerns of the urban century: Increasing agency and contested empowerment. In J. van der Heijden, H. Bulkeley, & C. Certomà (Eds.), *Urban climate politics: Agency and empowerment* (pp. 1–20). Cambridge University Press.

Visconti, C. (2017). Community-based adaptation measures for Water Sensitive Urban Design in context of socio-environmental vulnerability, *TECHNE: Journal of Technology for Architecture and Environment*, *14*, 352–361. https://doi.org/10.13128/Techne-20802

About the author

Gilda Berruti is an architect and Associate Professor of Urban Planning at the University of Naples Federico II, Italy. Her research interests include the social construction of urban plans, the sustainable city as an aspect of the new urban question and the relationship between planning rules and urban informalities.

Maria Federica Palestino is a Professor of Urban Planning at the University of Naples Federico II, Italy. She works on urban fragility through inclusive action-research strategies aimed at enhancing mutual learning and shared knowledge in urban planning, programs and policies.

7. Narratives on Babylon Hill

Exploring the making of a community and its urban forest through oral and environmental history (1985–2015)

Lise Sedrez and Natasha Augusto Barbosa

Abstract
Morro da Babilônia is an informal settlement located in Rio de Janeiro. Sedrez and Barbosa study a reforestation project implemented in that area through a collaboration between local residents and the municipal government. The authors employ archival research and oral history to recover the histories of the settlement and explore the evolution of the reforestation project. Throughout the 20th century, the memory of the residents discloses an ambiguous relationship with the forest and urban environment, marked by frequent unnatural disasters, mostly slides and floods. This story is not only about the transformation of a favela but also about the transformation of the actors involved in the reforestation project, all of them struggling with their preconceptions on humans and nature.

Keywords: reforestation, floods, grassroots and municipal collaborations, oral history, Rio de Janeiro

Back then, and I was one of those people myself, we didn't like people in reforestation projects. I didn't want people from the community getting in and out in the reforestation area.
—Luiz Lourenço, forest engineer, Municipal Secretary of Conservation and Environment, 2019

There is a forest at the top of Morro da Babilônia. It is a special forest, the result of a reforestation "collective effort," or *mutirão*. Someone put that forest there, and someone keeps nurturing the forest, with care, attention, science

Armiero, M., E. Turhan & S. Paolo De Rosa (eds), *Urban Movements and Climate Change: Loss, Damage and Radical Adaptation*. Taylor & Francis Group 2023
DOI: 10.5117/9789463726665_CH07

and dreams. The Morro da Babilônia forest is entwined in the landscape that features on one of the most celebrated postcards of Rio de Janeiro: Copacabana and the Sugar Loaf hill. The forest is present in the technical reports of the government agencies that designed the city's environmental policies. It is perceived in the region's microclimate, with thriving new—and yet centuries old—Atlantic Forest biodiversity. Finally, it is also recognized in the memory of the residents of Morro da Babilônia who planted, cared for and lived with the forest.

To study the relationships between society and nature in urban space, urban environmental historians need to interrogate the landscape, the archives and biodiversity, but we must also interrogate the memory that is created about the urban forest *by* the creators of that urban forest. This memory is fundamental to understanding the interaction of the local community with the biophysical space because, as we argue, it is through this interaction that the parameters of the community's right to the city are established—as well as its right to urban nature.

An oral history of the urban forest

Morro da Babilônia, located in Copacabana, Rio de Janeiro, has its own history of settlement, growth and community creativity. The origin of the current community dates to the 19th century; it was perhaps an urban *quilombo* like so many others, where makeshift shacks replaced stretches of sumptuous Atlantic Forest. Throughout the 20th century, the memory of the residents discloses an ambiguous relationship with the forest and urban environment, of acceptance and rejection, and eventually, of a successful experience of reforestation.

The reforestation project is recent. The first effective steps, as we will see, date to the late 20th century, in the 1980s. This is a history of the present time, but also a history of trees, of precarious housing, of low-priority policies, of struggles of mid-level municipal agents and a history of the trajectory of Rio de Janeiro in recent years. It is, thus, an ideal case for combining oral history and environmental history. For a discipline that insists that social relationships are inseparable from the biophysical world, environmental history has not engaged often with the promises and challenges that oral history brings, with some honorable exceptions. In 2004, in a roundtable at the annual meeting of the European Society for Environmental History, Jan Oosthoek (2015) summarized how he saw the role of oral history in environmental history: basically unexplored. Ten years would pass before the Canadian journal *Oral History/Forum d'histoire orale,* published a special issue on environmental history (Campbell, 2010). More directly, two important works in recent years explored the connections between oral history and environmental history.

The first, an edited volume organized by Katie Holmes and Heather Goodall (2017), brings together works by environmental historians who use oral history to construct complex and rich narratives about communities in Asia, Africa and Oceania. The chapters dialogue with postcolonial studies and highlight the role of local communities in challenging traditional narratives of the colonial past about place and practices in the biophysical world. Holmes and Goodall underline the ways in which oral history questions traditional concepts such as pollution, good living and conservation. In this perspective, communities that were forcibly removed from polluted areas in Bangladesh, for example, increased the pressure on urban green areas and recalled nostalgically the rivers where they once lived and that were defined as polluted by the municipality. The authors' goal is not, of course, to suggest greater tolerance for pollution, but to argue that the relations of place and community, of the right to urban space, are more complex than environmental indicators allow.

The second work is an article by Brian Williams and Mark Riley (2020), in which the authors address the connections between environmental history and oral history from a more theoretical perspective. The authors argue that oral history can contribute to environmental history from three important perspectives: knowledge, practices and power. Obviously interconnected, the three perspectives allow us to understand some of the potential of oral history in the construction of narratives about the environment and society. For knowledge, the authors suggest that oral history reveals a memory of past environments by the communities who occupied these places of study; by practices, how these communities worked and lived in a dynamic environment, and finally, for power, the inequality of access to natural resources and the conflicts in the disputes over these resources.

These two works helped us to frame our research in the community of Morro da Babilônia. They suggest, in terms of methodology, guidelines of encouragement and caution, which we have diligently used in our work. We consider three particularly important steps, drawn from these guidelines. First, the importance of contextualizing our oral sources and substantiating them with other sources. Oral history does not replace archives, but rather offers new lenses for examining them. Thus, we delved into newspapers and diaries in search of news about Morro da Babilônia, its origins and its nature. The bytes of the digitized archives have replaced the dust of the old and romantic newspaper libraries, but the search for these voices from the past remains fundamental. Second, oral history transforms perspective on place, a concept dear to environmental history, by demanding an active presence.[1] Momentarily leaving behind the archives, we climbed the Morro da Babilônia for our interviews. We went through the forest paths and alleys between houses and the replanted forest to come up with our interview structure. We saw *boa constrictors*—new residents attracted by the reforestation project—not far from areas with more than a hundred years of occupation, at the top of the Morro, where once thatched houses stood—until they gave way to brick and concrete, although always marked by the precariousness of housing. Finally, as Holmes and Goodall (2017) correctly point out, we adopted flexibility as our principle regarding the types of interviews we conducted. Some were individual conversations; others, technical meetings or focus groups, and, still others, collective interviews with local workers. As interviewers, we adapted to the interviewees, not the other way around. Thus, this text compares collective interviews of the oldest residents of the community with archival research, mostly contemporary newspaper articles. We also listened to the workers of the reforestation project, a joint initiative by the Rio de Janeiro city hall, the private sector and civil society, about their perceptions of the urban nature in Morro da Babilônia and their expectations about the meaning of the project for the community.

In this research, we looked at records of the earlier occupation of Morro da Babilônia in archives and libraries. According to our research, the military was already there in the 18th century, through the establishment of fortifications meant to protect the city from possible invasions. According to Ana Carolina Botelho (2016), there was an artillery battery on the slope of Leme with a path to Botafogo during the period of the Marquis de Lavradio's government (1769–1778). This military unit was expanded in 1822, when the Forte do Leme was built in the same place.[2]

"People of the worst kind"

The urban transformations that took place in the city of Rio de Janeiro, with their aspirations to public health and modernity, also led to a renewed occupation of the Morro da Babilônia. The dwellings of the urban poor

1 For a discussion of place in environmental and oral history, see Endres, 2011.

2 Another fort, currently named Forte Duque de Caxias, was built on the top of Forte do Leme from 1913 to 1919.

in Rio de Janeiro were largely slums and shacks in the hills, spaces with little infrastructure, but affordable. The celebrated urban reforms in the downtown sought to eliminate them from the noblest part of the city, in the same movement that made housing informality viable in the outskirts. Favelas such as Morro da Babilônia, not too close to downtown, but also not too far, in up-and-coming neighborhoods like Copacabana, blossomed and became symbols of the attempt to eradicate the historical slums (Abreu & Vaz, 1991).

In fact, on June 2, 1907, Morro da Babilônia made the headlines of the *Correio da Manhã* newspaper, alarmed by its rapid settlement. The article contrasted the beauty of the landscape with the misery that was climbing the hill. The journalist portrayed the misery as "poor homeless people, who were exposed to the weather, expelled from the huts in which they lived by edicts of progress, and that gradually occupied the hill. The progress of the modern city where old paths, once imprisoned by the hollows, now spread out into roads, in streets where electric trams glide." On Morro da Babilônia, the journalist met the former private soldier from the Army Corps of Engineers, José Carlos de Andrade, who, since he could not afford the living costs in the city, "has retreated to the mountain." He said that he had been on the hill for 14 years, that used to sell canes "down here," and that when his vegetable garden did well, his partner would go downhill to sell produce in Copacabana.[3]

Newspapers wrote about Morro da Babilônia as if it were the synthesis of many urban problems gathered in just one place, the "favela," still in quotation marks, by analogy to the slum in Morro da Favela, not yet a word on its own as it would become later in the century. It was an imaginary type presented as the antithesis of the so-called formal city. Thus, on December 16, 1915, another article in *Correio da Manhã* described the "*favelados*" of Morro da Babilônia as people of the "worst kind" who would spoil the hygiene and aesthetics of the beautiful spot that had attracted many for its luxurious green cover.

> The residents of Leme are condemned to see their properties damaged and their lives threatened without a measure being taken by those who are responsible for looking after them and guaranteeing them … an agglomeration of hovels built of kerosene cans and boxes, inhabited by people of the worst kind and without profession, thus harming the

3 *Correio da Manhã*, June 2, 1907, p. 1.

conditions of hygiene and salubrity of the place, and the aesthetics of that corner, so sought after due to the exuberance of your vegetation.[4]

Fifty years later, the communities of those hills, including Morro da Babilônia, were back in the headlines as places of risk. Not only the risk of diseases and insalubrity, as in the news article above, but also of landslides due to torrential rains. The 1966 rains were some of the worst storms in Rio de Janeiro's history (Sedrez & Maia, 2011). But it was not exactly the rains that constituted the disasters. Heavy rains are part of the history of Rio de Janeiro. We must place tragedies involving natural phenomena within a context of differentiation of the urban space in terms of socio-environmental vulnerability. In other words, it was not rain that caused the precarious housing situation that scared the residents of Morro da Babilônia, but political, urban and historical decisions. The socio-environmental disaster, therefore, was a process that began long before the rains fell on the city (Sedrez, 2013).

This scenario of hills crashing over the city in dramatic landslides, a powerful image that marked the collective memory of Cariocas, empowered the mobilization for public actions to remove the favelas and recover the slopes. According to reports published in the newspaper, *O Globo* (April 1, 1969; September 18, 1969), the Geotechnical Institute, created after the rains of 1966–1967, began containment of the city's slopes, both with structural works and with a plan to prevent erosion and landslides caused by rains. In 1969, simultaneously, the institute began reforesting an area of 340,000 square meters, including certain parts of Morro da Babilônia. For this project, some residents were removed to housing projects in Cidade de Deus and Vila Kennedy, on the other side of the city. Interviewed by the newspapers, the displaced families expressed their concern "for the problems they will face, living so far from their jobs and Copacabana beach."[5]

The process initiated by the Geotechnical Institute in 1969 was a harbinger of what would become the Mutirão Reforestation Program in the 1980s. According to the Municipal Environment Department, the initiative for the Mutirão program emerged out the public authorities' concerns regarding the improvement of the quality of life of the marginals who occupied irregular areas, such as the hills. So, the Municipal Department of Social Development of Rio de Janeiro began the Mutirão program, which aimed at the urbanization of communities with local labor. In 1986, the increase in the number of slums and the danger of landslides led to the inclusion

4 *Correio da Manhã*, December 16, 1915, p. 2.

5 *O Globo*, October 7, 1969, p. 12.

of reforestation plans in the initial project. In 1994, the Reforestation Task Force was transferred to the Environmental Recovery Coordination Office of the Municipal Secretariat of the Environment (SMAC), which increased the number of work fronts.[6]

Remembering socio-environmental change

The Mutirão Project began in Morro da Babilônia on July 28, 1995, with the planting of 21,600 seedlings. Shortly after, in 2001, a public lawsuit forced Shopping Center Rio Sul, convicted for degrading a protected area to expand its parking lot, to sign a term of adjustment of conduct (TAC) with residents' associations. The TAC's goal was the recovery and environmental preservation of the hills of Babilônia and São João, through the implementation and maintenance of reforested areas by SMAC. This is the context of the origins of the Mutirão de Reflorestamento project in Morro da Babilônia, and what we wanted to revisit in our interviews. Our main objective was not to obtain new data about the project, elements that archival records could have missed, but we hoped to understand how those involved in the project (residents and civil servants) remembered it as a foundational factor in the relationship between the community and the biophysical world.

For this purpose, the methodology of oral history was crucial, mostly for its challenges to the concepts of memory and historical past. The connections between history and memory imply a certain partition of what we call the past. History entails intelligibility and reflection on the historical processes, while memory is a subjective mechanism that is based on collective or individual reports and that contributes new perspectives on past events. Thus, the latter becomes the domain of investigating the former, as Enzo Traverso (2012, pp. 22–23) argued. Memory, always engendered by the present, through oral sources, gives us access to personal files that reside between remembrance and oblivion, claims Lucília Almeida Neves Delgado (2010, pp. 20–30). More directly related to our research, as Holmes and Goodall (2017) argue, oral history underlines the importance of memory "for the stories that ordinary men and women tell about environments in which they move or cross."

As an attempt to record these narrated stories, these "personal files" not yet written, we present in this chapter some excerpts from interviews conducted with eight residents of Morro da Babilônia, and with two

6 Available at: www.rio.rj.gov.br/web/smac/recuperacao-ambiental.

municipal technical civil servants who participated in the initial project at the then Municipal Secretariat of the Environment (SMAC) of Rio de Janeiro. The interviews about Morro da Babilônia took place through an informal network established with the help of the president of the reforestation cooperative, CoopBabilônia, Carlos Antônio, better known as Palô. During our preliminary contact, we revealed all the guidelines and objectives of our project, as well as the relevance of the residents' testimonies. From then on, we agreed that the interviews would take place with women from the community, in a conversation circle, therefore as collective interviews. The three interviewers were also women, and we hoped to establish trust by using a traditionally feminine space, a kitchen. We had no opportunity to talk to the informants beforehand, and so we were very eager to build trust as well as we could. In January 2019, when we held our meeting, the community was still reeling from the violence that had taken place in May of the previous year during a dispute between two rival criminal groups. The memory of alliances and betrayals from that period was still fresh, and the coexistence between the community's diverse sectors was tense. Our presence as interviewers was an unknown element in an already complex labyrinth.

Morro da Babilônia is divided into two favelas, Babilônia and Chapéu Mangueira. The latter is called Chapéu Mangueira (Mangueira Hat) because of plans to establish a hat factory there in the 1940s, plans which never took place. A large landmark with the name of the future factory is all that remains of those plans. Today it refers indeed to the name of the place and also indicates the boundary between the two communities. Their long coexistence has been marked by moments of solidarity and estrangement. The last few years have been a period of almost complete separation. We arrived at the Chapéu Mangueira sports court, which at the time of the interview was hosting a summer camp for children from both communities.[7] The summer camp, in fact, was one of the few joint activities shared by Babilônia and Chapéu Mangueira. We met in the kitchen, as our informants engaged in various activities for the summer camp. Some of the women were from Chapéu Mangueira, and others were from Babilônia. A total of six women agreed to share with us their knowledge and perspectives on the Mutirão Project during their work break from the summer camp activities. Two other women initially chose not to participate, but as the conversation progressed, they also contributed comments and suggestions.

7 Six women, collective interview, January 23, 2019. Their names were changed as requested.

This interview design, almost an informal roundtable, helped to defuse some initial resistance. The informants could participate or remain silent as they wished. At the same time, they helped each other to remember stories. In cases of environmental history, this collective remembering draws from the experience of sharing the same biophysical space, and not so much from the direct knowledge of our object of study, the Mutirão Project. Indeed, only one of the informants had participated directly in the Mutirão Project, and she was interviewed again later, individually. For this collective interview, however, our objective was to understand the relationship of the residents with the dynamics of the biophysical environment and its transformations over time.

A forest of memories

The Occupy Climate Change! project, which inspired our work, focuses on climate issues and community activities. However, concepts such as microclimate or global warming are not necessarily part of the everyday vocabulary of the Carioca population. Thus we decided to begin by asking questions about how much value they attribute to the forest near their houses. Their perceptions regarding the forest and the benefits of reforestation turned out to be closely related to their experiences of the climate, to the well-being that one has on the forested hill as compared to the hot weather in other neighborhoods, such as in this excerpt:

> Sandra: Yes, it is important.... Oxygen, the cool air. Wow, you come back [from work], take a bus and go to the North Zone where there are few trees, when you come back, you take a bus through the Aterro do Flamengo, you already feel like, wow, you're back to civilization. You breathe, the air is different (laughter).
>
> Luiza: It's different.
>
> Sandra: It's different, it's different, it's different. Regardless of whether it's hot here, in Caxias [a North Zone municipality] it's worse, much worse.

Later, we asked again about local natural phenomena such as erosion and rainfall. We asked about specific rains they remembered. The testimonies then referred to a strong rainstorm in 1989, when a landslide caused fatalities. The date, critical for the history of the community, was also marked in the body of one of the interviewees. She remembers the date well because she was in her son's postpartum period. Another informant also recalled that "she narrowly escaped being swept away in the landslide, since she was supposed to have had coffee with her son's godmother, whose house was swept away by the torrent." The community's social relationships—pregnancy, visiting neighbors, frustrated expectations—marked the memory of socio-environmental events and the way they intertwined the biophysical world and the social world.

> Interviewer: And those houses that fell? Were they right in the center, right in the center of the community, or were they closer to the woods?
>
> Luiza: They were up there, above the association headquarters.
>
> Interviewer: Way up the hill by the association headquarters?
>
> Maria: If I had gone for that coffee with the girl, I would have [suffered the same fate] too.... I went to visit her [in the hospital]. She had the [dressing].... She had shaved her head, she was there because she had broken her neck, which they shouldn't have moved because only the firefighters could move [a body], but the firefighters were late, then people went to help, right? I mean, the way they got her they must have broken her spine. Then she stayed at [Hospital] Souza Aguiar.... [She had] wounds that they didn't even clean. I only know that she suffered [a lot]. Then she even told me—she called me "Paraíba"—"Paraíba, I was making a cake for us to have coffee. I don't know what happened since then. I only know that the cake was in the oven." And thus happened the first disaster here. I remember.
>
> Interviewer: Are we talking about 1989? Could it have been 1994?
>
> Maria: No, 1989, right?
>
> Paula: No, it was in 1989 when my son was born. I was in the postpartum period.[8]

8 There was another landslide in the community in 1994, which, however, was not recalled by the informants. The interviewer sought to corroborate the dates, and later, we confirmed that in June 1989 there was indeed a landslide with fatalities in Morro da Babilônia. Six women, collective interview, January 23, 2019. Their names were changed as requested.

Of the six women we interviewed, only Maria, 78, was born and raised in Morro da Babilônia. Her memories, as well as our archival research, register the presence of military activities. Maria said that in "the time of the military" it was not easy to build a house. The army exercised their power over the region, and they blocked the construction of shacks and hovels. By inference, we concluded that the army deterred not only the expansion of individual buildings, but the expansion of the favela's built area.

> Maria: I remember that when I was little, the army was in charge here. So, we couldn't make houses. They were adobe houses, but we couldn't make them, enlarge [the already existent ones], nothing. If we did, they would come and knock them down. So we had to live with a bunch of kids in a tiny little room, and that was it. Got it?

This topic, regulating the growth of the Morro da Babilônia favela, was a point of contact for another interview, which we carried out in the same month, January 2019, with two technical civil servants of the Municipal Secretariat for Conservation and the Environment (Seconserma). Here, too, the place of the interview is relevant. Instead of a kitchen, we met in a room in the City Hall of Rio de Janeiro that housed the Mutirão Project for decades: paperwork, personnel, pictures and archives. Cláudia Ribeiro França and Luiz Carlos Pereira Lourenço are both forest engineers who participated in the project from its very beginning. According to França and Lourenço, the city government initially thought of implementing reforestation in the hills with the main goal of controlling the expansion of slums. As Lourenço said, "If I plant [trees] uphill from this community,... the community will not grow [in that direction] because there is a forest there."[9]

This interview also revealed that the idea of reforestation arose only in the second stage of the Mutirão Project. The original project was created by the Secretariat of Social Assistance. As it got underway, communities were called upon to provide the labor for urbanization and local improvements, while the municipality offered technical guidance. This was how the informants described the beginning of the Mutirão Project:

> Luiz Lourenço: So, let's see! The Mutirão, not necessarily the program, this Mutirão Project was created in '86, right Cláudia?... In the former Municipal Department of Social Assistance, which today is the Municipal Department of Social Assistance and Human Rights. Yeah, it was created

9 Claudia França and Luiz Lourenço, interview, February 8, 2019.

out of the premise, a disadvantaged community in which people build things on their own. They built stairways, sewers, to improve the community. But what happened? You're making the stairway, but when the stairs reach your own house, it is done. It is no longer my problem. The city government understood that if it started to pay for the task force, it could set the goals for it. [It could assure] conditions for the services to go on, with a minimum of productivity, quality, everything. Then the city hall began to provide materials, and [contacted] the residents' associations, and everything else, got to know the people [involved in the works] and the city hall technicians gave technical support. "Look, the stairways have to be built like this" or "Sewage pipes have to be placed like this."

Claudia França: Daycare, too.

Luiz Lourenço: Daycare, too, right? And so it goes. And back in '86, I think someone had the idea, the great idea, why not include reforestation? Not because back then anyone had an environmental vision. It was like this, if I plant [trees] uphill from this community,... the community will not grow [in that direction] because there is a forest there, and it is easier to monitor [the expansion of the community over the forest] than if it were just grass. And I'm going to create a security structure for the community, because I will have a forest there that will increase infiltration, decrease surface water runoff, that is, instead of the water running all at once, it will run little by little.

According to both city hall officers, their relationship with the community had been based on dialogue, trying to establish planting sites that were good for reforestation and for the community. They hoped to get the attention of residents so that they would participate, get involved and value the Mutirão Project. Their technical perspective, however, did not always match the expectations of the residents, especially about what to plant.

The questions for this kind of interview were necessarily different. We had built trust with the residents in our joint conversation, through questions about shared memories, about the experience of living in that biophysical environment, and we let the informants themselves define the vocabulary of the interview. With the technical personnel, we built trust using other kinds of shared vocabulary, one that values their technical training. The environmental historian's questions called for memories of what happened, but they also evoked the informant's professional expertise.

Our earlier questions gave them the opportunity to expand on their own experience:

> Interviewer: So, let me ask you some more technical things. An urban forest is a planned forest. You use the things you have in the place, but ecological functions are also expected, since this is a forest that will coexist with society. What are your criteria in a project for an urban forest?

At the same time, we were open to the possibility that their expert opinions at the moment of the interview might differ from what they thought in the past they were recalling because expertise changes over time. Our informants might have learned from their remembered experience, and our questions valued their learning process.

Negotiations and transformative engagements

In his responses, Lourenço revealed some of the Mutirão Project's challenges and the transformation in his own professional journey. The project could take into account only one type of knowledge, environmental engineering, but it was also a political project, obligated to consider the community as a partner. Lourenço needed to find a balance for his dual role: forest engineer and city hall employee.

> Luiz Lourenço: The first step is, you must use labor from the community. In the case of the reforestation *mutirão*, we needed to establish a forest uphill, above the location of a community. Because [this is] the only legal labor justification, to be able to use the *mutirão*, people who don't have a formal contract, they don't have anything [labor rights]. So, I can't do it [plant the forest] wherever I want to. That's the first step. In addition, we always need the residents' association to contact us [first]. As much as I would like to work in an area, the request must come from the association, you know? It's even a matter of us supporting the residents' association.

Hence this text's epigraph, which expresses the transformation of a professional. If initially he saw the community as a "necessary evil," maybe even a hindrance to the execution of an eminently technical work, over time as he worked for the project, his perspective changed, and the community became a treasured partner. To preserve and foster this partnership, it was even worth it to make the project more flexible.

> Luiz Lourenço: Back in the day, and I was one of those people. I particularly didn't like having [nontechnical] people in the reforestation [area]. I didn't want people from the community to enter the reforestation [area].
>
> Interviewer: Let us do our work, [you thought]....
>
> Luiz Lourenço: Don't step on my seedlings! (laughs). You're getting in the way! "I would like you to plant jackfruits, I would like you to plant oranges...."
>
> Claudia França: Yes, that's what they were always asking! For us to plant fruit trees....
>
> Luiz Lourenço: Traditional fruit trees. And we have evolved. [Today] we plant fruit trees, but I don't do everything they want. Today I can see it with different eyes. I was trying to think, how can I win people over to the project, you know? My idea was to "buy" people from the *mutirão*, to "sell" them the idea of the forest and have them take it to the community. To let them realize that the climate has improved, that there are fewer landslides,... but that they must leave the forest be, they must not come to disturb it. Because it's already a hard, arduous job, to carry seedlings, to cut grass.... It's complicated. People pass by, they step on a seedling. I know it, and the people in the fieldwork know how much work it took to plant that seedling there. Today I can see a little further, I can try. Depending on the situation, if what it takes is for us to make an orchard, we do try.
>
> After all, the forest was worth an extra jackfruit tree.

França and Lourenço spoke of the Mutirão Project with great pride as a project that has survived multiple municipal administrations. Despite the conflicts that arise in the interdependent relationship between residents and technicians, and even taking into account the thorny political disputes over project maintenance and funding, the Mutirão Reflorestamento is a successful project. França and Lourenço listed for us the Mutirão Project's awards: it was selected for the Mega-Cities Project, sponsored by the United Nations; it was listed among the "100 Brazilian Experiences of Sustainable Development" (1990); it won the Environment Prize of the Regional Council of Engineering and Architecture of Rio de Janeiro (1998); the Model Project of the Society for Ecological Restoration (1999); Honorable Mention at the Metropolis Awards in Seoul South Korea (2002), and several others. In fact, they attributed the project's longevity precisely to this international recognition. Few mayors would dare to shut down such a renowned and award-winning project.

França and Lourenço also underlined how the communities learned to appreciate the benefits brought by the project such as a pleasant microclimate. At the same time, they recognized that what first interested those communities was the very modest remuneration the workers received for participation—although, over time, since the project has lasted over three decades, that initial interest has turned into passionate advocacy for the project, and there is no fault in that, they claim.

> Interviewer: [Can we say that] that most requests [for reforestation] from residents' associations were [made] because they wanted [job opportunities, that they were] thinking about employment, about work?
>
> Cláudia França: Yes, yes! That's the reason....
>
> Luiz Lourenço: It is a source of income for the people. Over time, we managed to work with this, and today, the people of the Mutirão, there are the people who defend [it]. [They] beat their chests and say, "I did it!" The guy defends [the forest]. If someone tries to cut down trees, the guy goes there [and says,] ... You won't build [houses] here, you won't!
>
> Cláudia França: Involvement comes over time!
>
> Luiz Lourenço: "Here I helped to create [it], you won't destroy this here!" [In] some places it doesn't work, the community doesn't work very well, but there are places where it works!

Lourenço and França recognized that there are many challenges involved in keeping full activities going in a 30-year-old project, challenges such as the violence and criminality that has sometimes threatened civil servants and foresters employed by the city hall, but they were convinced of the need to ensure community participation when promoting any kind of work in favelas.

> Luiz Lourenço: I won't hire a private company to work there.... [You must] have reforestation through the community,... [otherwise] it doesn't work.
>
> Interviewer: Did you do that?

Luiz Lourenço: We have tried....

Claudia França: Even hiring local labor.

Luiz Lourenço: Even hiring local labor, it doesn't work, because then you must pay "tolls," and you have to do that....

Interviewer: "Tolls," what are talking about?...

Luiz Lourenço: Criminal faction tolls, understand? You have to.... [You'd think it is easy,] you'd think "the public power has already penetrated the community, it has already created Favela Bairro there."[10] First, [Favela Bairro] is a gigantic, huge package. It is not a reforestation action, it is a package that covers [services for] the entire community, so it's very intense and you go, you do it and you leave. Not the Mutirão.... The reforestation project ... is going to enter [the favela]. It is going to plant [the trees], and you have to maintain it. You have to go there to take care of it, so you create [something with] a very long lifespan that creates a complicated cycle.... So, it doesn't work to enter with a private company.

Luiz Lourenço: Yes, away from the favela, we do have companies that [we work with], they have already done technical research, they have already done estimates, in areas without favelas, we can do it. Now, inside a favela community, it's very complicated.

These two models of interviews allow us to compare how the involved actors see not only their own actions in the biophysical environment, but also the relationships they have established with each other. Community residents and city hall technicians have different perceptions of their actions on Morro da Babilônia, even when they work in the same space and with the same objective. Reforestation, to be successful, depends on both the engagement of residents, who see the forest's benefits for their community (such as the microclimate and protection against landslides) and on the professional dedication of city hall technicians. Environmental historians, armed with the resources of oral history, seek to identify both the engagement of the first and the professionalism of the latter. With that, we observe the transformation of the residents of the favela of Morro da Babilônia, who came to value the

10 Favela Bairro was an public urbanization project that aimed to provide public services to the favelas in Rio de Janeiro and was initiated in 1995.

forest they planted; of the city hall technicians, who came to appreciate their partnership with the community where they work; and of the biophysical environment, which now has a new forest covering it.

The transformations of residents, technicians and the biophysical world are part of a process that not only reaffirms the *favelados'* right to the city, but also their right to urban nature, a dynamic nature, in continuous transformation, but fundamental to understanding the experience of urban living.

Acknowledgements

This chapter is part of the Occupy Climate Change! project, associated with the Environmental Humanities Lab, KTH, Stockholm, Sweden. Research for this chapter was funded by the Swedish Research Council for Sustainable Development (FORMAS) through the National Research Program on Climate (Contract: 2017-01962_3). An earlier version was published as Sedrez, L. F., & Barbosa, N. A. (2019). Narrativas na Babilônia: Uma experiência de história oral, risco climático, reflorestamento e comunidade (1985–2015). In Andrea Casa Nova Maia. (Ed.), *História oral e direito à cidade* (pp. 79–98). Letra e Voz.

References

Abreu, M. A., & Vaz, L. F. (1991). Sobre as origens da favela. In *Anais do IV Encontro Nacional da ANPUR: Velhas e Novas Legitimidades na Reestruturação do Território* (pp. 481–492). ANPUR.

Botelho, A. C. B. (2016). *Os fortes como esquinas da cidade: O uso público no Forte Duque de Caxias, Leme, Rio de Janeiro.* PhD thesis, Universidade Federal do Rio de Janeiro.

Campbell, C. E. (2010). "We all aspired to be woodsy": Tracing environmental awareness at a boys' camp. *Oral History/Forum d'histoire orale, 30.*

Castro, A. H. F. (2009). *Muralhas de pedra, canhões de bronze e homens de ferro: Fortificações do Brasil de 1504 a 2006.* Fundação Cultural Exército Brasileiro.

Delgado, L. A. N. (2010). *História Oral: Memória, tempo e identidades.* Autêntica.

Endres, D. (2011). Environmental oral history, *Environmental Communication, 5*(4), 485–498. https://doi.org/10.1080/17524032.2011.610810

Holmes, K. & Goodall, H. (Eds.). (2017). *Telling environmental histories: Intersections of memory, narrative and environment.* Palgrave Macmillan.

Oosthoek, K. J. (2015, July 27). The role of oral history in environmental history. *Environmental History Resources*. https://www.eh-resources.org/role-of-oral-history-in-environmental-history/

Sedrez, L. (2013). Desastres socioambientais, políticas públicas e memória: Contribuições para história ambiental. In E. S. Nodari & S. M. Correia (Eds.), *Migrações e natureza* (pp. 185–202). Oikos.

Sedrez, L., & Maia, A. C. N. (2011). Narrativas de um dilúvio carioca: Memória e natureza na grande enchente de 1966. *História Oral, 14*(2), 221–254. https://doi.org/10.51880/ho.v14i2.239

Traverso, E. (2012). *O passado, modos de usar: História, memória e política*. Edições Unipop.

Williams, B., & Riley, M. (2020). The challenge of oral history to environmental history. *Environment and History, 26*(2), 207–231. https://doi.org/10.3197/096734018X15254461646503

About the authors

Lise Sedrez is Professor of Environmental History at the Universidade Federal do Rio de Janeiro, Brazil. Her work has been published in Italy, Colombia, Brazil and the USA. She is a research scholar for the Conselho Nacional de Desenvolvimento Científico e Tecnológico (CNPq) and a postdoc fellow at the Universidade Federal de Minas Gerais. Her research interests include urban environmental history, global environmental governance and history of disasters.

Natasha Augusto Barbosa is an environmental historian and a PhD student at Casa de Oswaldo Cruz, Fiocruz, Brazil. Her research interests include urban socio-environmental relations with a focus on community action to build healthy and sustainable urban environments, oral history and public history.

8. Repositioning marginal spaces in climate adaptation

Periphery, power and possibility

Karen Paiva Henrique

Abstract

A growing body of literature has exposed how efforts to convert cities into climate-safe zones intertwine with patterns of uneven development, spatial segregation and profound inequality, often reinforcing marginalization of subaltern communities. Henrique proposes an empirically rich and theoretically sophisticated case study on São Paulo (Brazil). In 2009, the local government announced the removal of 7,500 low-income families along the Tietê River to implement a multi-million-dollar adaptation plan aimed to protect the city from floods. Adopting a feminist approach, the author examines the grassroots resistance to this and other government interventions, showing how the coproduction of spaces, bodies, and political subjects can provide lessons for more just adaptation efforts within and outside the floodplain.

Keywords: mitigation, eviction, political subjectification, São Paulo, urban margins

> *This is an intervention. A message from that space in the margin that is a site of creativity and power, that inclusive space where we recover ourselves, where we move in solidarity to erase the category colonized/colonizer. Marginality as site of resistance. Enter that space. Let us meet there. Enter that space. We greet you as liberators.*
>
> —bell hooks (2015, p. 234)

Cities are at the center of global climate change and adaptation debates for several interconnected reasons. First, the fast-paced development of urban

Armiero, M., E. Turhan & S. Paolo De Rosa (eds), *Urban Movements and Climate Change: Loss, Damage and Radical Adaptation*. Taylor & Francis Group 2023
DOI: 10.5117/9789463726665_CH08

centers and their elites' unsustainable consumption patterns (Dodman, 2009; Satterthwaite, 2008) contribute to global emissions that lead to irreversible climate change (Revi et al., 2014; Rosenzweig et al., 2010). Second, and due to changes in climatic patterns, cities are hotspots for both extreme and slow-onset events (Revi et al., 2014) that disproportionately impact poor groups (O'Brien, 2017). Third, cities are often considered, by adaptation scholars and practitioners alike, "beacons of hope" (Watts, 2017, p. 537), operating as innovation hubs to curtail emissions and promote timely and effective adaptation (Castán Broto & Bulkeley, 2013). As city governments embrace the role of first responders to climate change (Rosenzweig et al., 2010; Watts, 2017), they increasingly incorporate concepts such as sustainability, resilience and smart urbanism in their planning lexicon; however, not always to everyone's benefit (see Derickson, 2017; Fainstein, 2015; MacKinnon & Derickson, 2013). A growing body of literature exposes how efforts to convert cities into climate-safe zones entwines with historical patterns of uneven development, spatial segregation and profound inequality, further marginalizing those who are already most vulnerable (Anguelovski et al., 2016; Shi et al., 2016).

For example, in the face of exacerbated floods, the government of São Paulo is implementing a multi-million-dollar adaptation plan, *Parque Várzeas do Tietê* (PVT, or Tietê Lowlands Park), a 75-kilometer-long floodplain restoration scheme that combines new infrastructure with plans to remove and relocate informal settlements and allegedly protect São Paulo against future flooding (Departamento de Águas e Energia Elétrica, 2015). Under development for over a decade, the PVT has had a series of adverse effects on the communities it directly affects, including heightened financial burdens and environmental risks for displaced families as well as exacerbated levels of anxiety and insecurity for those whose future remains uncertain (for details, see Henrique & Tschakert, 2019a). These negative effects notwithstanding, the project continues to move forward. Its proponents also insist on branding low-income floodplain dwellers as both environmental villains and recurrent flood victims while simultaneously rendering their voices, needs and desires invisible (Henrique & Tschakert, 2019b). It is precisely this erasure-in-action that this chapter aims to confront.

To this end, I focus on the margins of São Paulo's urban development, the city's East Zone, henceforth also referred to as the periphery, centering my analysis on the Tietê River floodplain and its subjects. While margins and peripheries are sometimes distinguished within critical urban geography as geographical locations and alternative modes of urbanization that can "happen anywhere" (Caldeira, 2017), respectively, here I reconcile the two concepts because São Paulo's East Zone can be read simultaneously as both. Drawing on the theoretical and methodological approaches advanced by feminist scholars attentive to marginal and embodied city dynamics, and encounters with community leaders, I examine how the coproduction of spaces, bodies and political subjects can provide lessons for more just adaptation efforts within and outside the floodplain. Ultimately, this contribution aims to locate marginal voices at the center of climate adaptation discussions. It does so by making visible the ways that the urban periphery matters to those who inhabit it and how their knowledge and experience as political agents can inform more hopeful trajectories toward uncertain climate futures.

Feminist approaches to margins, subjectivities and bodies

Feminist scholars, including, but not limited to, critical (urban) geographers and political ecologists, have long challenged abstract and disembodied notions of the global. On the one hand, they highlight how global processes are experienced in daily practices (Elmhirst, 2011; Truelove, 2011). On the other, they demonstrate how these intimate experiences produce social, economic and environmental relationships that often transcend the spaces of the everyday (Mountz & Hyndman, 2006; Tschakert, 2012). As these feminist scholars "ground, locate, map and link empirical realities" of living and knowing the global (Mountz & Hyndman, 2006, p. 458), they expose broader politico-economic forces' perverse effects on marginalized citizens while revealing how these forces are contested, resisted and reworked in and through marginal landscapes.

This feminist attention to marginal spaces and relationships invites adaptation scholars to examine the situated and embodied effects of global climate change as it intersects with other social, economic and political forces. It also provides entry points to redefine climate adaptation based on the lived experiences, knowledges and actions of those who are disproportionately affected by its obliterating forces. Here, I build on diverse and rich contributions from feminist scholars to position the urban margin, its subjects and their bodies at the center of the analysis and rearticulate climate adaptation from the ground up.

Repositioning margin as center

The urban margin has long been critically examined as the spatial counterpart of broader politico-economic systems and their associated unequal resource distributions. It is conceived as the outcome of a state planning apparatus that responds to capitalist market demands by etching informality and difference into the urban fabric (Baviskar, 2003; Roy, 2009), producing an uneven landscape of vulnerabilities and risks. Following this capitalist logic, poor people, people of color and ethnic minorities are increasingly confined in urban "misery belts" (Lugo-Vivas, 2020), i.e., spaces where accumulated layers of disadvantage translate into a perpetual struggle over rights and resources. As a result, the urban margin remains paradoxically inside the formal city, indispensable for its reproduction (see Rolnik, 2007), yet perpetually on its outside, configuring spaces where marginalized residents are denied access to the opportunities commonly associated with urban life.

Even so, feminist scholars contend, the margin should not be approached only as a site of permanent lack. For example, contributions from Black and Chicana feminists highlight how the margin should also be appreciated as a transitional space, as a hybrid landscape of "wounding, healing and then empowerment," and as a site of resistance (Anzaldúa, 1987; Mountz & Hyndman, 2006, p. 452). Seen through this lens, the margin is conceived simultaneously as a site of systemic deprivation and of "radical openness" (hooks, 2015, p. 223) that is the epicenter of counter-hegemonic strategies to produce the urban through words, habits and everyday practices (hooks, 1984, 2015).

Examining subjectivities in and through space

A feminist reading of margin as center invites a more nuanced understanding of peripheral subjects. A long research tradition in feminist scholarship, which includes contributions from cultural geographers and political ecologists, has focused on how identity constructions and representations are deeply connected with the spaces where they are produced (Longhurst, 2003). In other words, "to be is to be somewhere" (Longhurst, 2003, p. 5), and people's subject positions, or subjectivities, are produced through multiple and intersecting relations of power that are "performed, resisted, disciplined and oppressed not simply in, but through space" (Brown & Knopp, 2003, p. 11).

This conceptualization calls attention to how space informs subjectivities, and subjectivities rearticulate the definition of space (Probyn, 2003) in fluid relationships where neither becomes ever fixed or stable (see Valentine, 2007). Through this lens, the margin, conceived as a site of radical possibility (hooks, 2015) perpetually in-the-making, is made visible as a space of encounters and action where subjectivities are forged with emancipatory potential for both marginalized citizens and spaces.

Focusing on the body as political space

Subjectivities are not, however, only constituted through geographical space; they are also produced in and through the body (Longhurst, 2003), which only becomes whole in relation to its social and physical context (Grosz, 1998, p. 43). As the subject's first and most intimate environment (Rocheleau et al., 1996), the body mediates one's experiences in the production of both subjectivities and space. In this context, the city is understood as the "place where the body is representationally reexplored, transformed, contested, reinscribed" (Grosz, 1998, p. 47) along multiple, intersecting axes of difference (e.g., class, gender and race) (Doshi, 2017; Heynen, 2018). Urban spaces produce different bodies through corporeal exertion and cultural saturation (Grosz, 1998). The body, in turn, transforms the city according to its evolving needs, procuring resources in the spaces of the everyday (Truelove, 2011). This search for resources through individual everyday practices and collective action then affects the body, reinscribing or transforming subjectivities in relation to the materiality of space (Nightingale, 2011; Sultana, 2009). This reciprocal production of cities and bodies calls attention to the ways in which marginal urban spaces produce and reproduce bodies that are subjugated by a permanent lack of resources, and in which marginalized bodies reclaim urban space in return.

Such an embodied reappropriation of space informs the production of urban subjects in at least two interconnected ways. First, it engenders intimate ways of knowing the city through "material knowledges," i.e., knowledges produced through contact, touch and taste that allow one to know *with* rather than *about* something or somewhere, thus offering more grounded ways of knowing the world (Porto-Gonçalves, 2017). Second, the embodied appropriation of space generates deep emotional connections to place and a sense of belonging (Manzi et al., 2018) as individuals with different aspirations and skills come together to produce and protect the spaces they have reclaimed.

Feminist attention to corporeal experiences makes visible how bodies become intimate sites for the production of knowledge, emotions and history (Probyn, 2003), which then translate into political action that often transcends their everyday spaces (Nightingale, 2012). Such action involves a number of practices that have the potential to confront ubiquitous neoliberal development agendas (Doshi, 2017; Elmhirst, 2015) and, I argue, can inform transformational adaptation built from the bottom-up, based on negotiation, recognition, trust and respect, also in the context of climate change.

In this chapter, I employ a feminist lens to focus on the margins of neoliberal urban development and articulate a counternarrative against hegemonic disembodied practices and modes of representation produced by the state. In the process, I illustrate how the embodied production of space becomes entwined with the production of political subjects, and to what effects. Repositioning the margins, I argue, can reveal what people value and why, and what they do about it to enrich "understandings of the terrain of political transformation" (Doshi, 2017, p. 127; see also Heynen, 2018) and inform more just adaptation futures.

Stories from São Paulo's eastern periphery

In the 1970s, the city of São Paulo started to expand along the Tietê River to accommodate a thriving industrial complex and its ever-growing low-wage workforce (Zanirato, 2011). Lacking access to capital, and thus São Paulo's more affluent and environmentally safer neighborhoods, these predominantly Black and Brown migrant populations settled in low-lying areas without land tenure, infrastructure and services, and that are particularly susceptible to flooding. Over the years, informal settlements continued to expand largely without state investments, configuring landscapes of risk, vulnerability and persistent cycles of land dispossession. Underpinning these dynamics are deeply ingrained value systems that privilege the interests of powerful actors such as the industries that play a key role in environmental degradation and the production of risk in the basin (see Henrique & Tschakert 2019a) over the lives and livelihoods of historically marginalized and oppressed communities (see Henrique & Tschakert, 2021).

It is in this context that projects intended to adapt São Paulo to future flooding are envisioned and must be understood. In 2009, the government of São Paulo announced the removal of 7,500 low-income families from the Tietê River's eastern periphery to implement the PVT (Departamento de Águas e Energia Elétrica, 2015). To justify these removals, the state mobilized denigrating portrayals of informal floodplain communities by depicting them simultaneously as environmental villains who pollute the river and clog drainage networks, and as victims adversely impacted by its recurrent overflow (Henrique & Tschakert, 2019b).

In the summer of 2009–2010, the state government was able to capitalize on the city's worst flooding season in contemporary history and expedite the removal of thousands of families from the floodplain, the majority of whom were displaced without a definitive housing solution (Henrique & Tschakert, 2019a). Yet, this was by no means an unencumbered process. Floodplain residents have employed a myriad of resistance tactics to thwart forced removals (Barboza, 2015; see Henrique & Tschakert, 2019a), animated by an unwavering commitment to fight over the right to live dignified lives, have a voice and be treated with respect.

At the center of this fight is Ronaldo *do* Pantanal (Ronaldo *of* Pantanal),[1] a local community leader who has devoted his life to building and maintaining the neighborhood of Jardim Pantanal. Over the years, the trajectories of Jardim Pantanal and Ronaldo have become deeply entangled, giving rise to an emblematic political figure devoted to the fight for his community's right to city space. Here, I focus on significant moments of this shared trajectory to unravel the complex relationship between the production of the periphery and the political subject and to examine its implications for the continuous expansion of São Paulo's neoliberal urban form in a changing climate. I draw inspiration from Valentine (2007) and Moser and Law (1999) to narrate this trajectory in seven stories based on several encounters with Ronaldo and other community leaders between February and June 2017. Following these and other feminist scholars (see Sundberg, 2004), I carefully recreate these stories as they happened. Yet, the research and the narratives that follow are still filtered through my experiences as a white, middle-class, Brazilian woman who grew up in Brazil yet was fundamentally an outsider to São Paulo's periphery and thus relied significantly on key informants to gain access. Navigating such tensions allowed me to examine how spaces and political subjectivities are produced, reinforced and/or contested even in the process of doing research.

The stories do not always describe an independent event, nor do they follow a chronological order. All the interactions described in this chapter took place in Portuguese, my native language. The interviews with Ronaldo and the other two community leaders featured in the stories, Zélia and Vagner, were audio recorded and transcribed. All the other activities, including participant observation, transect walks and informal interactions, were

1 In Portuguese, the preposition "*do*" (as in Ronaldo *do* Pantanal) indicates both that Ronaldo is *from* Pantanal (that he is from there) and that he is *of* Pantanal (that he belongs to that community). Here, I am emphasizing the latter meaning.

recorded in fieldwork diaries. Together, the recordings and the diaries provide a rich dataset from which to derive the detailed accounts presented below. As the participants' requested, all the names mentioned in this chapter are real.

Story 1

I first met Ronaldo outside of *Instituto Alana*, a nonprofit organization in Jardim Pantanal. I had first contacted Ronaldo by email at the suggestion of other participants who were familiar with his work in Pantanal. He promptly replied and invited me to meet him in the community. My initial encounter with the neighborhood was through the windows of an Uber I had caught at the local train station. The driver knew better than to follow the GPS on his smartphone and in a few minutes delivered me at the institute's doorsteps.

There, I met Ronaldo, an enthusiastic man in his 50s, who smiled broadly as he talked to residents surrounding him on all sides. Sporting a short brim hat and a red button with the slogan, *Fora Temer!* (Out with Temer!, then president of Brazil), he warmly welcomed me and invited me to join him for lunch at a friend's *boteco* (a small local restaurant) a short walk from the institute.

On our way to lunch, we talked for the first time in detail about my doctoral project. He could tell from my southern accent that I wasn't from São Paulo. I could tell from his northeastern accent that he was not from São Paulo either. Even so, he said he had lived in Pantanal for decades, since the early days of the community. The short walk from *Instituto Alana* to the *boteco* turned out to be much longer than I expected. Long walks over short distances would become the mark of all my visits to Pantanal in Ronaldo's company. After every few steps, Ronaldo stopped to greet someone or, more often, was stopped by someone who wanted to say hi and ask how he had been doing. I would soon learn that Ronaldo no longer lived in São Paulo (he had relocated to another city due to political animosities and personal threats), but returned at every chance he had to participate in community activities and visit friends.

As people stopped him along the way, Ronaldo got caught up in conversations because he, too, wanted to know how his long-time neighbors were doing in his absence. A five-minute walk turned into a 30-minute walk. With his arms stretched wide open, he spoke happily and loudly to everyone who approached him. Ronaldo was visibly feeling at home. As he reiterated at every new introduction I would witness, Ronaldo was *of* Pantanal.

Story 2

I met Ronaldo again at another local train station where we were joined by a group of undergraduate architecture students and their instructor for a guided visit through Pantanal. As we left the station, Ronaldo and I started chatting again. After having failed to identify the exact perimeter of Pantanal on official maps, I had decided to ask him for the boundaries of the community. We started to walk toward the Tietê River, when I asked him: "Ronaldo, where does Pantanal begin?" Ronaldo stopped. Pointing to the line where the asphalt ended and the street continued as a dirt road, he looked at me and said: "You asked me where Pantanal begins? Pantanal begins here, where the asphalt ends."

We resumed our visit, getting closer and closer to the river. Ronaldo stopped every few meters, pointing at different places and sharing information about the community. He called attention to residents' measures to mitigate floods and described how the state failed to provide for the community. Ronaldo told the group there was a time when Pantanal did not have piped water and residents were forced to make *gatos* (illegal connections) to access potable water inside their homes. Ronaldo then described how he and other residents organized a *mutirão* (a cooperative) and installed water pipes for 39 streets in the community. Soon after, the community was approached by representatives from SABESP (the state water management company) who wanted to connect the new pipes to water meters so that the residents could be properly charged for the water they were consuming. With a voice that exuded both resentment and determination, Ronaldo explained how the residents told SABESP they could install the water meters *only* if they provided sewage treatment for the entire community.

Story 3

As the sun soared above us, indicating that it was getting closer to midday, we headed to *Instituto Alana*, the final stop of our visit. Suddenly, the instructor rushed to our side with a piece of paper in his hand. As he moved closer, I noticed the paper was a printed map of Jardim Pantanal. Pointing to the map, the instructor asked Ronaldo to take us down a certain street before we went back to the institute. Apparently this was the street where the students would develop their projects that term. Ronaldo explained to the instructor that the street did not exist. The instructor insisted the street had to exist because it was on the map. Ronaldo later explained to me that this was a perfect example of how the people deciding the future of Pantanal (including architecture students who later become city planners) are too far removed from the communities where they intervene. According to Ronaldo:

> They come here, and they are blind. They ... can't find themselves here.... Then, when they come to the community and you explain, as a guy who lives here, who designed it, who developed it, who built it with his own hands, they just don't believe you.

Story 4

Ronaldo invited me to meet him at a university situated close to the city center. As soon as his other commitments were over, we walked to the cafeteria to chat about Pantanal. As we sat down, Ronaldo pulled a laptop from his bag and started to go over a detailed presentation about the history of the community. Most slides focused on the 2009–2010 floods and the evictions the government enacted that summer to begin construction of the PVT.

Some slides showed pictures of residents moving along flooded streets, carrying their belongings, while the caption read "Desperate residents ... try to salvage whatever they can." Others showed men hired by the government to destroy homes after their inhabitants had either been forcefully evicted or had fled to escape the prolonged floods. Ronaldo shared with me the conversation he had with one of the men demolishing the houses, who was a local resident and had accepted the job to provide for his family. Yet, he also knew that someone else would be hired to do the same to his home, which had also been built in an informal part of Pantanal. As he said this, Ronaldo recalled, the man started crying.

Ronaldo continued to show his slides, arguing that the government was invested in removing informal residents from Pantanal and used the PVT to justify forced evictions. However, as Ronaldo proudly explained, rather than remaining silent, the community protested systemic neglect and indiscriminate evictions. He showed pictures of large groups of residents being violently beaten and pepper-sprayed, but still standing their ground against police forces.

Then he shared images of the projects he and other floodplain residents had developed with universities over the years to legalize the Pantanal. All of them involved local labor and capacity building, which Ronaldo considers central to any inclusive effort for Pantanal's development. He also explained that none of these solutions were ever embraced by the government because they would not have turned a profit for São Paulo's construction industry. Ultimately, he concluded, "the community doesn't have a voice."

Story 5

My next visit to Pantanal took place on a cold Saturday morning. Ronaldo was giving a lecture to a group of students from the community. I arrived early and watched as the large classroom was quickly filled. Ronaldo introduced me to the group. I described my research project and sat in the back, listening and taking notes.

Ronaldo asked the students: "How many of you have ever experienced a flood?" At least 15 immediately raised their hands. Ronaldo then proceeded to show a series of slides about Pantanal; these were the same slides he had showed me a few weeks earlier. My experience seeing the presentation in the community classroom, however, was entirely new. I watched as his words were followed closely by the mesmerized students, who expressed surprise about how little they knew about their own community.

Ronaldo discussed how the floods in Pantanal are exacerbated by private companies and developers who operate unencumbered within and outside the floodplain, as well as by the state, which refuses to invest in local infrastructure and maintenance. In response, a student uttered in disbelief, "I have always been told that flooding is caused by informal communities alone!" Ronaldo then talked about community action, including the development of infrastructure, and concluded with a video of the protests he and other residents organized during the floods of 2009–2010.

As soon as Ronaldo's lecture was over, he and the students invited me to join the *fórum estudantil*, a weekly meeting where young community members discuss broad social issues. That morning's topic was "*A luta das mulheres*" ("The women's fight"). I watched as everyone passionately joined the debate. The female students used the space to share personal stories of discrimination and challenge gender roles. The male students at times insisted on women's responsibility for maintaining the household, but still engaged respectfully in the discussion.

The students went on to discuss final arrangements for other forum activities and details about a march they were organizing to protest Michel Temer's presidency. The students talked about creating posters for the protest while more senior members shared strategies for dealing with the burning sensation caused by the pepper spray often deployed by the police. I noticed as a woman in her mid-30s joined the conversation.

After the meeting, Ronaldo introduced me to her. Her name is Zélia, and she is a leader from Chácara Três Meninas, an informal community that neighbors Pantanal. I asked her if she would participate in an interview for my research. She said yes, adding that she was only agreeing because Ronaldo trusted me: she was tired of people she did not know twisting her words.

Story 6

I met Zélia outside a public school in Chácara Três Meninas, and we went for a walk in the neighborhood. She was eager to discuss the relationship between informal settlements and the river, and explained how private companies were destroying the environment by encroaching on and polluting the Tietê, but remained unchallenged by the PVT. After a 30-minute walk around the neighborhood, we came across a bench on the sidewalk, and we sat down and continued our conversation there. As with Ronaldo, our chat was interrupted by her neighbors who wanted to ask how she was doing. I asked Zélia if she could tell me a bit more about herself.

Zélia had migrated from Bahia and, upon arriving in São Paulo, was extremely shocked by the level of inequality that plagued the region:

> There [in Bahia] we shared everything we had.... When I arrived here [in São Paulo], and I saw this inequality, homeless people, people begging, especially kids ... without school, without anything, I couldn't accept it.... I say that this place has been the best university of my life ... because this is exclusion.... We feel totally excluded. [We feel] like nobody, like rubbish, like the worst radioactive rubbish you can imagine, one that no one wants to be close to.

She explained that this feeling of being utterly marginalized contributed to her desire to participate more in the community, but added that it was Ronaldo who persuaded her to take on a leadership position. She described how she was attending a meeting about the provision of electricity in the community when Ronaldo suddenly stood up and said: "And now, a new leader from Chácara Três Meninas will speak!" Smiling, Zélia recalled that she quickly left the room without saying a word. Soon after, however, she became actively involved in fighting for the rights of her community. She concluded:

> The most interesting thing is that I found a space not only to speak, but a space where I could act! There is a lot to be done, but also other things

we have already accomplished. Not a whole lot, but it already makes a difference.

Story 7

In another visit to Jardim Pantanal, Ronaldo introduced me to Vagner, a man in his mid-30s who lived close to *Instituto Alana*. Vagner grew up around the community center, where he spent all day "jumping around on the sand and picking fights with everyone." As the years passed, the community center became a space where he learned and matured. Vagner explained: "We see things happening, see the leaders intervening with the residents, creating relationships through these interventions.... It brings you automatically closer to the space."

Vagner described how he initially only observed community activities, many of them organized by Ronaldo, and then started to take part in the action, including several protests during the floods of 2009–2010. According to Vagner, together the residents guaranteed several improvements for the community, including paved streets, electricity, piped water and zip codes, but political disagreements eventually corroded the leadership, and the movement lost traction.

Around the same time, Vagner started working with marginalized youth from other peripheral neighborhoods who have no opportunities to rise above their constrained livelihoods. Vagner tries to help them with daily tasks and expose them to different spaces and cultural activities. He added: "Education is a shot in the dark, you don't know where it is going to lead. You contribute what you can, you plant the seed. Whether they take it or not is up to them."

The embodied coproduction of space and political subjects

As I interacted with Ronaldo and other community leaders over five months of fieldwork, it became evident how residents, even if not originally *from* São Paulo's peripheral communities (i.e., having been born elsewhere), become *of* that community. That sense of belonging is attached to their bodies and subjectivities, transcending the community's physical space. Such embodied sense of belonging was clear in the case of Ronaldo, who had left Pantanal, but whose attachment to that space defines how he continues to position himself in relation to his community, transpiring, for example, in the way he still introduces himself as Ronaldo *of* Pantanal.

Ronaldo's sense of belonging is not only the result of having lived in Pantanal for decades, but also, and perhaps more importantly, of having invested years of physical and emotional labor to build a livable community in the absence of a liable state. It is a sense of belonging that comes from individual and collective efforts exerted through the body, producing sweat, sore muscles, calluses and bruises, as residents work with little resources to build their everyday spaces from the ground up. This sense of belonging fuels and is ignited by a fight over recognition, as residents insist on their right to own the communities they have produced.

Attention to the relationship between marginalized bodies and the spaces they inhabit, articulated through this persistent fight, reveals the intricate and contentious ways flood adaptation plays out in the Tietê River floodplain, as well as the multiple and contested subjectivities such processes engender. On the one hand, the floodplain dwellers' bodies bear the brunt of recurrent floods aggravated by a historic resource deficit, which they continue to compensate for as they expand and consolidate their communities, pushing the margins of informal development toward the Tietê. The state then exploits these communities' disproportionate exposure and rampant expansion to justify adaptation efforts that replace them with an exclusionary archetype of environmental preservation, epitomized by the PVT. On the other hand, residents fight back against the project as yet another attempt of incorporation by a seemingly opportunistic state, using their bodies as spaces for resistance and as receptacles of values, knowledge and experiences that are transmitted to others within their communities. In the process, floodplain residents subvert state representations that brand them simultaneously as villains of environmental degradation and as victims of the floods, repositioning themselves as political subjects.

These political subjectivities then translate into action, as detailed in Ronaldo's accounts of community resistance and negotiations with state officials for the provision of infrastructure. Such situated politics is also evident as Ronaldo welcomes university researchers and students into the community, replacing preconceived notions of the periphery with a situated narrative carefully crafted by the floodplain residents themselves. Yet, reappropriating floodplain narratives is not an easy task as residents struggle to have their local knowledges recognized even by seeming allies, a role here embodied by the instructor. This is not to say that the instructor questioned Ronaldo's familiarity with his space out of malice, but rather to highlight the extent of a systemic marginalization that places residents' material knowledges outside of formal (and thus authoritative) epistemic frameworks.

Reshaping dominant narratives in and through the periphery is central to the constitution of political subjectivities that are always in the making, as demonstrated by Ronaldo's lecture and other activities in the student forum, where younger residents become invested in the history of the community through exercises of collective remembering. Here, I borrow from hooks (2015, p. 226) to argue how these activities constitute an act of "politicization of memory that distinguishes nostalgia, that longing for something to be as once was, as a kind of useless act, from that remembering that serves to illuminate and transform the present," and which thus holds great transformational potential. In Jardim Pantanal, this potential is realized as everyday community spaces become knowledge-sharing centers where senior residents keep local histories and values alive, training a new generation of engaged citizens. Such knowledges include situated material responses to everyday flood threats as well as practical resistance tactics shared with young activists, which they then apply to their own organized actions against incorporation strategies imposed by the state.

The significance of spatial and embodied political relations is also highlighted in the mobilization of new political leaders, such as Zélia and Vagner, who emerge within and sustain existing leadership networks, acting on a shared sense of responsibility and care for the periphery and its people. Moreover, engaged citizenship often transcends the space in which it is nurtured. This is perhaps most visible in Vagner's commitment to work with marginalized youth from other communities, connecting different peripheries in the pursuit for a better life.

This is not to say that the political subjectivities I encountered in the floodplain are unproblematic. In fact, as floodplain dwellers organize to secure resources and rights, they also exclude other residents whose values and desires they disregard as incompatible with their own. In my research, such exclusions were most salient when community leaders suggested whose accounts were worthy (or not) of appearing in the research. Even so, the stories related in this chapter weave together a situated narrative of struggle, unwavering commitment and profound attachment to place that demonstrates how political subjectivities are forged within and through marginal urban spaces.

Transposed to the context of climate change, these lessons offer strategic openings to reframe adaptation according to what people value most. They challenge the assumption that relocating the urban poor, the state's "go to" solution (see Henrique & Tschakert, 2019b), is the only and/or best answer to São Paulo's flooding problem by revealing that the floodplain is not only a space of persistent struggle, but also of belonging, pride and collective history. Recentering adaptation discussions around what marginalized residents value invites a progressive politics that builds on situated capacities and committed local leadership to negotiate possible solutions and inevitable trade-offs. The stories presented here reveal opportunities to foster such progressive politics and empower local leaders as community–government liaisons, nurture existing political spaces for youth engagement, and draw on everyday practices and situated proposals to pursue community adaptation in place.

Looking out from the margins within

As global environmental change scholarship recognizes adaptation's unjust implications, researchers redirect their attention to the question: "Adaptation for whom, by whom and how?" (Anguelovski et al., 2016, p. 333). Underlying the answer to this question are the institutional legacies and associated power hierarchies that determine whose values, experiences and knowledge count in the development of adaptation programs. Although scholarship on climate change and cities has grown notably in the past years, there is still urgent work to be done to achieve just and effective adaptation (Dodman et al., 2019). This chapter argues that such an endeavor will benefit from feminist engagements on at least two important fronts.

First, using a feminist lens allows us to not only investigate the uneven effects of global environmental threats, but also the intimate experiences of urban planning approaches designed to cope with such risks and harms. This lens makes visible how adaptation becomes entwined with uneven development trajectories that incorporate peripheral urban space and displace its informal residents for profit. These interventions often evade public records and survive only in the bodies of marginalized inhabitants, which are central to the multiple strands of feminist analyses presented here. Moreover, attention to the coproduction of spaces, subjectivities and bodies demonstrates how the periphery becomes the space in relation to which residents position themselves and others, highlighting yet again the profound and perverse effects of indiscriminate eviction programs.

Second, a feminist approach attentive to marginal and embodied city dynamics provides entry points to unlock the periphery's transformational potential. Just climate action, critical scholars argue, requires concerted efforts to transform adaptation from top-down, economically driven solutions into negotiation spaces where differently positioned actors work together to protect what matters most to them (Harris et al., 2018; Ziervogel et al., 2017; Barnett et al., 2016). By employing a feminist lens to reposition the margin as center, this article reveals how the periphery becomes the catalyst for the production of engaged political subjects who work upon a sense of individual and collective belonging to coproduce a shared vision for their communities.

Contesting master narratives of environmental change and their associated erasures-in-action requires "fostering spaces of listening, deliberation, debate, respect, imagination and trust" to reveal productive tensions and mediate differences (Veland et al., 2018, p. 45). This can be achieved through alternative modes of governance that embrace debate, struggle and everyday micropolitics in all their messiness (Nightingale, 2018), and cultivate resident–government negotiations from below (Ziervogel, 2019). A case in point, *Planos de Bairro*, that were developed in Brazil's northeast, bring local residents together to produce a collective community vision while building capacities to inform sustained emancipatory action (Manzi et al., 2018). This chapter contributes to this discussion by identifying opportunities for inclusive governance through city–body relations and their associated attachment to place.

Working with marginalized citizens within the spaces they inhabit has been a transformational experience. It has populated otherwise largely abstract adaptation projects with real bodies, their daily struggles and intimate desires. It has also opened a door to embrace "the positivity of otherness" (Rose, 1993, p. 150), as my research participants redefine preconceived notions of the periphery to advance a more progressive and inclusive vision of what climate-changed cities could and should look like based on a situated politics of belonging.

References

Anguelovski, I., Shi, L., Chu, E., Gallagher, D., Goh, K., Lamb, Z., Reeve, K., & Teicher, H. (2016). Equity impacts of urban land use planning for climate adaptation: Critical perspectives from the Global North and South. *Journal of Planning Education and Research, 36*(3), 333–348. https://doi.org/10.1177/0739456X16645166

Anzaldúa, G. (1987). *Borderlands/La Frontera: The new mestiza*. Aunt Lute Books.

Barboza, L. da S. (2015). *A identidade do movimento por urbanização e legalização do pantanal da zona leste—SP: Esperança e desalento na luta contra o deslocamento populacional*. Scortecci.

Barnett, J., Tschakert, P., Head, L., & Adger, W. N. (2016). A science of loss, *Nature Climate Change, 6*(11), 976–978. https://doi.org/10.1038/nclimate3140

Baviskar, A. (2003). Between violence and desire: Space, power, and identity in the making of metropolitan Delhi. *International Social Science Journal, 55*(1), 89–98. https://doi.org/:10.1111/issj.12184

Brown, M., & Knopp, L. (2003). Queer cultural geographies—We're here! We're queer! We're over there, too! In K. Anderson, M. Domosh, S. Pile & N. Thrift (Eds.), *The handbook of cultural geography* (pp. 313–324). Sage.

Caldeira, T. P. (2017). Peripheral urbanization: Autoconstruction, transversal logics, and politics in cities of the Global South. *Environment and Planning D: Society and Space, 35*(1), 3–20. https://doi.org/10.1177/0263775816658479

Castán Broto, V., & Bulkeley, H. (2013). A survey of urban climate change experiments in 100 cities. *Global Environmental Change, 23*(1), 92–102. https://doi.org/10.1016/j.gloenvcha.2012.07.005

Departamento de Águas e Energia Elétrica. (2015, November 29). *Parque Várzeas do Tietê—O maior parque linear do mundo.* DAEE—Portal Do Departamento de Águas e Energia Elétrica. http://www.daee.sp.gov.br

Derickson, K. D. (2017). Urban geography III: Anthropocene urbanism. *Progress in Human Geography, 42*(3), 425–435. https://doi.org/10.1177/0309132516686012

Dodman, D. (2009). Blaming cities for climate change? An analysis of urban greenhouse gas emissions inventories. *Environment and Urbanization, 21*(1), 185–201. https://doi.org/10.1177/0956247809103016

Dodman, D., Archer, D., & Satterthwaite, D. (2019). Editorial: Responding to climate change in contexts of urban poverty and informality. *Environment and Urbanization, 31*(1), 3–12. https://doi.org/10.1177/0956247819830004

Doshi, S. (2017). Embodied urban political ecology: Five propositions. *Area, 49*(1), 125–128. https://doi.org/10.1111/area.12293

Elmhirst, R. (2011). Introducing new feminist political ecologies. *Geoforum, 42*(2), 129–132. https://doi.org/10.1016/j.geoforum.2011.01.006

Elmhirst, R. (2015). Feminist political ecology. In A. Coles, L. Gray, & J. Henshall Momsen (Eds.), *The Routledge handbook of gender and development* (pp. 58–66). Routledge.

Fainstein, S. (2015). Resilience and justice. *International Journal of Urban and Regional Research, 39*(1), 157–167. https://doi.org/10.1111/1468-2427.12186

Grosz, E. (1998). Bodies–Cities. In H. J. Nast & S. Pile (Eds.), *Places through the body* (pp. 42–51). Routledge.

Harris, L. M., Chu, E. K., & Ziervogel, G. (2018). Negotiated resilience. *Resilience, 6*(3), 196–214. https://doi.org/10.1080/21693293.2017.1353196

Henrique, K. P., & Tschakert, P. (2019a). Contested grounds: Adaptation to flooding and the politics of (in)visibility in São Paulo's eastern periphery. *Geoforum, 104*, 181–192. https://doi.org/10.1016/j.geoforum.2019.04.026

Henrique, K. P., & Tschakert, P. (2019b). Taming São Paulo's floods: Dominant discourses, exclusionary practices, and the complicity of the media, *Global Environmental Change, 58*, 1–12. https://doi.org/10.1016/j.gloenvcha.2019.101940

Henrique, K. P., & Tschakert, P. (2021). Pathways to urban transformation: From dispossession to climate justice. *Progress in Human Geography, 45*(5), 1169–1191. https://doi.org/10.1177/0309132520962856

Heynen, N. (2018). Urban political ecology III: The feminist and queer century. *Progress in Human Geography, 42*(3), 446–452. https://doi.org/10.1177/0309132517693336

hooks, bell. (1984). *Feminist theory from margin to center.* South End Press.

hooks, bell. (2015). *Yearning: Race, gender, and cultural politics* (2nd ed.). Routledge.

Longhurst, R. (2003). Placing subjectivities. In K. Anderson, M. Domosh, S. Pile & N. Thrift (Eds.), *The handbook of cultural geography* (pp. 283–389). Sage.

Lugo-Vivas, D. A. (2020). Racialized and identity-based inequalities as (new?) frontiers for academic discussion: Future agendas around land issues. *Journal of Latin American Geography, 19*(1), 233–245. https://doi.org/10.1353/lag.2020.0023

MacKinnon, D., & Derickson, K. D. (2013). From resilience to resourcefulness: A critique of resilience policy and activism. *Progress in Human Geography, 37*(2), 253–270. https://doi.org/10.1177/0309132512454775

Manzi, M., Figueiredo, G. C. dos S., Mourad, L. N., & Rebouças, T. de M. (2018). Neighbourhood planning and the right to the city: Confronting neoliberal state urban practices in Salvador, Brazil. *International Journal of Urban Sustainable Development, 10*(1), 1–15. https://doi.org/10.1080/19463138.2018.1433677

Moser, I., & Law, J. (1999). Good passages, bad passages. *Sociological Review, 47*(1), 196–219. https://doi.org/10.1111/j.1467-954X.1999.tb03489.x

Mountz, A., & Hyndman, J. (2006). Feminist approaches to the global intimate. *Women's Studies Quarterly, 34*(1/2), 446–463.

Nightingale, A. J. (2011). Bounding difference: Intersectionality and the material production of gender, caste, class and environment in Nepal. *Geoforum, 42*(2), 153–162. https://doi.org/10.1016/j.geoforum.2010.03.004

Nightingale, A. J. (2012). The embodiment of nature: Fishing, emotion, and the politics of environmental values. In E. Brady & P. Phemister (Eds.), *Human–environment relations* (pp. 135–147). Springer Netherlands.

Nightingale, A. J. (2018). The socioenvironmental state: Political authority, subjects, and transformative socionatural change in an uncertain world. *Environment and Planning E: Nature and Space, 1*(4), 688–711. https://doi.org/10.1177/2514848618816467

O'Brien, K. (2017). Climate change adaptation and social transformation. In D. Richardson, N. Castree, M. F. Goodchild, A. Lynn Kobayashi, W. Liu, & R. A. Marston (Eds.), *The international encyclopedia of geography* (pp. 1–8). John Wiley & Sons.

Porto-Gonçalves, C. W. (2017). De saberes e de territórios: Diversidade e emancipação a partir da experiência latino-americana. In V. do Carmo Cruz & e D. Araújo de Oliveira (Eds.), *Geografia e giro descolonial: Experiências, ideias e horizontes de renovação do pensamento crítico* (pp. 37–54). Letra Capital.

Probyn, E. (2003). The spatial imperative of subjectivity. In K. Anderson, M. Domosh, S. Pile & N. Thrift (Eds.), *The handbook of cultural geography* (pp. 290–299). Sage.

Revi, A., Satterthwaite, D. E., Aragón-Durand, F., Corfee-Morlot, J., Kiunsi, R. B. R., Pelling, M., Roberts, D. C., & Solecki, W. (2014). Urban areas. In *Climate change 2014: Impacts, adaptation, and vulnerability. Part A: Global and sectoral aspects. Contribution of Working Group II to the Fifth Assessment Report of the Intergovernmental Panel on Climate Change* (pp. 535–612). Cambridge University Press.

Roberts, C., & Connell, R. (2016). Feminist theory and the Global South, *Feminist Theory, 17*(2), 135–140. https://doi.org/10.1177/1464700116645874

Rocheleau, D. E. W., Thomas-Slayter, B., & Wangari, E. (Eds.). (1996). *Feminist political ecology: Global issues and local experiences*. Routledge.

Rolnik, R. (2007). Territórios negros nas cidades brasileiras: Etnicidade e cidade em São Paulo e Rio de Janeiro. In R. Emerson dos Santos (Ed.), *Diversidade, espaço e relações étnico-raciais. O negro na geografia do Brasil* (pp. 75–90). Autêntica.

Rose, G. (1993). *Feminism and geography: The limits of geographical knowledge*. Polity.

Rosenzweig, C., Solecki, W., Hammer, S. A., & Mehrotra, S. (2010). Cities lead the way in climate-change action. *Nature, 467*(7318), 909–911. https://doi.org/10.1038/467909a

Roy, A. (2009). Strangely familiar: Planning and the worlds of insurgence and informality. *Planning Theory, 8*(1), 7–11. https://doi.org/10.1177/1473095208099294

Satterthwaite, D. (2008). Cities' contribution to global warming: Notes on the allocation of greenhouse gas emissions. *Environment and Urbanization, 20*(2), 539–549. https://doi.org/10.1177/0956247808096127

Shi, L., Chu, E., Anguelovski, I., Aylett, A., Debats, J., Goh, K., Schenk, T., Seto, K. C., Dodman, D., Roberts, D., Roberts, J. T., & VanDeveer, S. D. (2016). Roadmap towards justice in urban climate adaptation research. *Nature Climate Change, 6*(2), 131–137. https://doi.org/10.1038/nclimate2841

Sultana, F. (2009). Fluid lives: Subjectivities, gender and water in rural Bangladesh. *Gender, Place & Culture, 16*(4), 427–444. https://doi.org/:10.1080/09663690903003942

Sundberg, J. (2004). Identities in the making: Conservation, gender and race in the Maya biosphere reserve, Guatemala. *Gender, Place & Culture, 11*(1), 43–66. https://doi.org/10.1080/0966369042000188549

Truelove, Y. (2011). (Re-)conceptualizing water inequality in Delhi, India through a feminist political ecology framework, *Geoforum, 42*(2), 143–152. https://doi.org/10.1016/j.geoforum.2011.01.004

Tschakert, P. (2012). From impacts to embodied experiences: Tracing political ecology in climate change research. *Geografisk Tidsskrift/Danish Journal of Geography, 112*(2), 144–158. https://doi.org/10.1080/00167223.2012.741889
Valentine, G. (2007). Theorizing and researching intersectionality: A challenge for feminist geography. *Professional Geographer, 59*(1), 10–21. https://doi.org/10.1111/j.1467-9272.2007.00587.x
Veland, S., Scoville-Simonds, M., Gram-Hanssen, I., Schorre, A., El Khoury, A., Nordbø, M., Lynch, A., Hochachka, G., & Bjørkan, M. (2018). Narrative matters for sustainability: The transformative role of storytelling in realizing 1.5°C futures. *Current Opinion in Environmental Sustainability, 31*, 41–47. https://doi.org/10.1016/j.cosust.2017.12.005
Watts, M. (2017). Cities spearhead climate action. *Nature Climate Change*, 7(8), 537–538. https://doi.org/10.1038/nclimate3358
Zanirato, S. H. (2011). História da ocupação e das intervenções na várzea do rio Rietê. *Revista Crítica Histórica, Ano II*(4), 117–129. https://doi.org/10.28998/rchvl2n04.2011.0007
Ziervogel, G. (2019). Building transformative capacity for adaptation planning and implementation that works for the urban poor: Insights from South Africa. *Ambio, 48*(5), 494–506. https://doi.org/10.1007/s13280-018-1141-9
Ziervogel, G., Pelling, M., Cartwright, A., Chu, E., Deshpande, T., Harris, L., Hyams, K., Kaunda, J., Klaus, B., Michael, K., Pasquini, L., Pharoah, R., Rodina, L., Scott, D., & Zweig, P. (2017). Inserting rights and justice into urban resilience: A focus on everyday risk. *Environment and Urbanization, 29*(1), 123–138. https://doi.org/10.1177/0956247816686905

About the author

Karen Paiva Henrique is an Assistant Professor in the Department of Human Geography, Planning and International Development at the University of Amsterdam. Her research interests include the justice implications of urban climate adaptation and the emancipatory potential of critical visual methods and storytelling. She was a contributing author for the IPCC Special Report *Global Warming of 1.5°C* (SR15).

9. Immigrant communities in Europe as situated knowledge holders for postcolonial and feminist urban adaptation to climate health risks

Panagiota Kotsila

Abstract

While determinants such as race, ethnic background and migration status are increasingly considered to be key factors in vulnerability to climate health risk, explicit consideration of immigrants' climate health vulnerability in the European context is largely missing. In her contribution here, Panagiota Kotsila addresses this knowledge gap in the literature on climate change and migration studies by proposing a critical reconceptualization and planning for healthy and just cities. Building on critical studies of climate change adaptation and mobilizing postcolonial climate urbanism as an analytical lens, Kotsila argues for the importance of situated knowledges, feminist methodologies and migrant-centric epistemologies.

Keywords: climate urbanism, climate-related health impacts, migration, vulnerability, feminist methodologies

We need the power of modern critical theories of how meanings and bodies get made, not in order to deny meanings and bodies, but in order to build meanings and bodies that have a chance for life.

—Donna Haraway (1988)

Climate-related health impacts are increasingly being felt and anticipated across the globe (Smith et al., 2014). Meteorological and extreme weather events have been found to be the most common categories of climate impacts

---

Armiero, M., E. Turhan & S. Paolo De Rosa (eds), *Urban Movements and Climate Change: Loss, Damage and Radical Adaptation*. Taylor & Francis Group 2023
DOI: 10.5117/9789463726665_CH09

that affect health, with the most common climate health outcomes being respiratory, cardiovascular, neurological, infectious disease and mortality (Rocque et al., 2021). Adaptation to these impacts is emerging as of core policy importance, both for Europe and globally (EC, 2019, 2020; Marinucci et al., 2014; WHO, 2010, 2018), the focus often being on vulnerable regions and vulnerable people that are affected most. However, while race, ethnic background and migration are increasingly considered factors in vulnerability to such climate health risks in the analyses conducted by leading public health institutions, such as the Centers for Disease Control and Prevention in the US (Gamble et al., 2016, pp. 247–286), an explicit consideration of immigrants' climate health vulnerability in the European context is largely missing.

Many European cities are often purported to be beacons of sustainability and climate adaptation, albeit with contradictions in their equity and justice, while others are dealing with some of the most extreme impacts of climate change so far. At the same time, European cities constitute entry points and destinations of international migration, but have failed to become immigration-friendly places. I thus focus on Europe, not because the links between immigration and climate change elsewhere are less important or better studied, but mainly because Europe has been, especially over the last decade, a locus of international migration and of neocolonial politics of migration, whereby immigrants including refugees are sociopolitically excluded, economically exploited and subjected to violent border technologies and cultural racism. This combination of factors (climate risks, climate expertise and increased inequalities and exclusions affecting immigrants from Majority World countries) makes Europe a paradigmatic case for examining what and how can we learn about climate health risks starting from the experiences and knowledge of immigrants, situated as they are within current neocolonial politics of migration.

While research on the social determinants of health has paid attention to how factors such as socioeconomic background, employment conditions, access to social welfare and country of origin shape health outcomes, there is still limited knowledge about how immigrant communities are experiencing climate change pressures and related health impacts, and what knowledge and strategies they mobilize to confront such experiences in the destinations of their migration. Knowing that immigrants are a vulnerable population experiencing, in some aspects, discrimination and hardship similar to the socially weakest national population groups (Hemminki, 2014), the current context of increasing need for climate preparedness makes it timely and urgent to assess how immigrants' climate health vulnerability is shaped and confronted, both systemically and in everyday life.

As I will argue in the following section, part of the reason for this gap in knowledge on the nexus of climate change and immigration is, on the one hand, a dominating focus on the impact of climate change and migration in places of origin (Adger et al., 2018). On the other hand, when we look at the literature on immigrants in post-migration contexts, there is a diffuse sense of immigrants as people in need with little to offer—especially in terms of knowledge and insights about key social challenges (such as climate change). At the same time, mainstream political discourse in Europe, like elsewhere, portrays immigration as a social "burden," a problem and a crisis. Instead, and in resistance to such narratives, I here propose that if we are to address the social exclusion, inequalities and disproportionate exposure to health risks that immigrants often face, critical scholarship should treat immigrants as equal knowledge-holders, with experience, expertise and situated—embodied, partial and accountable—knowledge(s), who can offer crucial insight when it comes to reconceptualizing and planning for healthy and just cities. It is this kind of knowledge that urban policymakers and societies have yet to acknowledge, explore and embrace. Focusing on the climate-related vulnerabilities that *follow* migration would provide novel insight on who forms part of the "population of concern" in climate urbanism and how this relates to social, environmental and climate justice. Shifting attention to what immigrants *do* know about climate health risks and how they respond to them in ways that enhance their adaptive capacities and more broadly that of urban regions is to challenge common monolithic and colonial narratives of immigrant groups as solely lacking resources and knowledge.

In this chapter, I bring together literature in political ecology, anthropology, climate change, public health, environmental justice and migration studies, drawing new pathways for the study of vulnerability to climate health impacts through a focus on immigrant communities in European urban areas. Building on and expanding an understanding of adaptation as a critical metaphor for change which addresses structural and underlying drivers of inequality, I echo calls for just and transformational climate urbanism (Bulkeley et al., 2014; Robin & Castán Broto, 2021; Shi et al., 2016). I thus call for a closer look at the ordinary practices and knowledges of immigrants, arguing that postcolonial climate urbanism needs to involve "unexpected" and "othered" actors, subjectivities and knowledges that exist *within* so-called global cities (Andreucci & Zografos, 2022). I thus forge a postcolonial approach that not only locates the infrastructural "South" in cities of the Global North (Robin & Castán Broto, 2021; Silver, 2019), but, more daringly, focuses on understandings and strategies that have traveled with immigrant communities and enriched, adapted to and been enacted in Europe.

I thus advocate for more research that asks how immigrants, and particularity those from Majority World countries, are positioned in relation to adaptation efforts and associated processes of capital accumulation in climate/health governance in European urban areas. Consequently, I also ask how do these migrants understand themselves within those processes, since this may determine how they engage in or resist such efforts (Eriksen et al., 2015; Kaika, 2017). I propose an epistemological engagement with immigrants' situated knowledges (Haraway, 1988) in relation to climate impacts on health and their known responses to them. Such an engagement would involve examining the knowledge around climate adaptation that is being exchanged, practiced and negotiated "in between" the Western city and migrants' places of origin (Lawhon et al., 2014). This would mean validating and elevating immigrant communities' diverse experiences and understandings of climate health risks, and the types of adaptation knowledge they share, transform and develop in response to these risks. Drawing from critical medical anthropology and feminist political ecology studies (Carroll, 2013; Jackson & Neely, 2015), I also apply a cultural lens to vulnerability analysis that pays attention to identity and knowledge, and the way they are negotiated and embodied in the everyday practices of immigrant groups, taking into account the tangible and intangible aspects of vulnerability. Responding to calls for situated political ecologies and intersectional feminism and attention to embodied experiences and situated knowledge in urban climate justice studies (Agyeman et al., 2016; Anguelovski et al., 2020; Doshi, 2017; Ranganathan & Bratman, 2021; Rocheleau, 2015), this chapter examines the everyday modalities through which ordinary people and specifically immigrants (with different origins, age, gender, income and employment sectors) strive to provide for a healthy and meaningful lives in cities and the limitations that they face (Lawhon et al., 2014).

Cities as centers of climate impacts, immigration and adaptation

Cities are especially vulnerable to climate health effects, and they also represent a crucial component of local adaptation responses to climate change (Watts et al., 2020a). Emerging climate-related health risks in European urban areas include heat-related illnesses during prolonged and intensified heat waves, increases in infectious diseases affected by seasonal rainfall and temperature patterns, and extreme flooding and droughts (Smith et al., 2014). Heat-related illness such as heat stroke, heat stress and mortality caused by heat are exacerbated by global warming and the higher frequency of unusually high temperatures (Åström et al., 2013; WHO, 2017). Globally, the period from 2015 to 2021 has been the hottest ever recorded, with recent heat waves and record high temperatures soaring in western North America, the Middle East, the Sahel and a number of Mediterranean countries (Haddad, 2021). Heat waves are lasting longer, and exceptionally hot days are getting hotter. Projections show that annual fatalities from extreme heat in Europe alone could rise from 2,700 deaths per year today, to up to 50,000 deaths per year by 2050 with a global warming range of 1.5°C to 2°C (EC, 2020). Cities are at high risk of these types of impacts due to the heat-island effect (Ramamurthy & Bou-Zeid, 2017). Urban populations in European cities are among the most vulnerable to extreme heat globally, with record-breaking temperatures and hundreds of heat-attributed deaths in western Europe (France and the Netherlands) in 2003, 2018 and 2019 (Watts et al., 2020b). In the US, research from 208 different cities showed that large increases in temperature will lead to excess mortality from heat events even in areas with high air conditioning coverage in place (Lay et al., 2021).

The climate is also rapidly becoming more suitable for infectious disease transmission due to shifting temperature and rainfall patterns, meaning that climate change expands the geographical risk of diseases to places and populations previously never exposed to them. The climatic suitability range for dengue, for example, a disease transmitted by mosquitoes, increased by 15% in 2018 (Watts et al., 2020a). West Nile virus (WNV), another disease transmitted by urban-dwelling mosquitoes, is seeing a global spread that is being accelerated by climate change among other factors (Paz, 2019; Smith et al., 2014) and has seen a significant increase in Europe in the last years. In 2018, Europe had 1,832 cases in total—more than seven times the number of cases than in 2017 and significantly higher than the case numbers for the previous 10 years—and remarkable outbreaks in 2020 in parts of Greece, Spain and the Netherlands (Bakonyi & Haussig, 2020).

Cities are a primary destination for international immigration, both globally and in Europe, mostly due to (expected) higher employment opportunities (Ramirez et al., 2018). Immigrants, especially those from Majority World countries, have limited access to health care, often due to their residence/citizenship status, their employment conditions, discrimination and the lack of culturally appropriate health systems in place (Benach et al., 2010; Castañeda et al., 2015; Lassetter & Callister, 2008; Lebano et al., 2020).

In cities, health inequities such as these are combined with unequal exposure to environmental health risks, which in turn are driven by patterns of racial and ethnic segregation or ghettoization in underserved neighborhoods (Wacquant et al., 2014). Immigrants, as well as people of color and ethnic minorities, are often more exposed to environmental contamination than other residents, while also being deprived of health-promoting amenities such as parks and accessible healthy/fresh food stores (Anguelovski, 2016; Bullard & Lewis, 1996; Heynen et al., 2006; Pulido, 2010; Ranganathan & Bratman, 2021). Such structural inequalities are also expected to shape exposure and vulnerability to climate health risks related to heat waves and/or infectious diseases, especially because immigrants tend to work long hours outdoors, are paid low salaries, and often live in overcrowded housing conditions of face housing deprivation (Hansen et al., 2013; Kotsila & Kallis, 2019). For heat wave-related health impacts, for example, we know that socioeconomic deprivation and poor housing are strong indicators of vulnerability (Macintyre et al., 2018). In the case of infectious diseases, such as dengue, too, social inequality and deprivation are primary vulnerability factors (MedECC, 2020; Watts et al., 2020a).

In order to confront emerging climate change impacts, cities around the world have been designing adaptation plans and policies, often under wider umbrellas of resilience and sustainability (Shokry et al., 2020; Woodruff et al., 2018). These efforts include the renaturalization of urban environments with the creation of new green/blue infrastructure (parks, gardens, waterways, etc.). Critical scholars have time and again noted how such efforts tend to prioritize a balance between economic goals and environmental sustainability and adaptation, but rarely include justice in their assessments, plans, processes and implementations (Agyeman & Evans, 2003; Connolly, 2018; Pearsall & Pierce, 2010). As a result, and despite their status as some of the most climate-affected groups, immigrants have limited access to the benefits of such efforts, precisely because of their minority status and misrecognition in political and civil life. This exclusion entails that their views, interests and representatives are often excluded from, or devalued in, decision-making processes regarding urban planning and adaptation (Anguelovski et al., 2019; Shi et al., 2016; Zografos et al., 2016).

Despite increased interest in immigrant health in Europe, considerations of race, ethnic background and migrant status have yet to be explicitly considered in assessments of climate health vulnerability and adaptation. This partly owes to the fact that, while most research on the interface of climate change and immigration focuses on migration as a potential adaptation strategy (Adger et al., 2018; IOM, 2015), little is known about the climate health exposures immigrants face and the adaptation strategies that they employ in migration destinations. In other words, while much of the literature has focused on out-migration as a response to climate change, there is not enough work on how immigrants experience, react to and shape responses to climate change in migration *destinations*, such as cities in the Global North. In the current economic crisis of more than a decade's duration, with increasing electoral support for far-right parties feeding on and reproducing xenophobia, this missing focus risks further alienating immigrants from the processes of city-making and avoiding responsibility for growing inequities in urban health and climate adaptation. Instead, as I argue here, including analysis of immigrants' vulnerabilities and adaptive capacities to climate change through the lens of political ecology will shed light on the structural and systemic drivers of such vulnerability, and create openings for progressive and transformative politics by demonstrating how, when immigrant rights and social welfare are supported, urban climate adaptation can be both more effective and more just.

Postcolonial climate urbanism through a focus on immigrant communities

Climate change adaptation, as a field of knowledge production and policy, has been criticized as a form of government technology, a governmentality that permeates day-to-day practice and the culture of everyday life (Stripple & Bulkeley, 2015). As Mills-Novoa et al. (2020) note, "adaptation projects mobilize distinct imaginaries and knowledge claims that create territories for intervention (the objects) as well as targeted populations (the subjects) to sustain them." Arguably a field heavily bounded by technocratic politics, critical geographers see in adaptation limited potential for socio-environmental transformation and justice (Taylor, 2014; Watts & Bohle, 2015). In cities, the growing consensus—if not impetus—for adaptation often builds on and reinforces asymmetrical power relations and disproportionately affects disadvantaged populations. This can be seen in the example of forced relocation into new urban/peri-urban developments under the pretext of adaptation and resilience to climate change impacts, as has been the case in São Paulo, Brazil (Millington, 2018), and urban regions of the Mekong Delta in Vietnam (Chun, 2015), among many others.

Urban environmental justice scholars and activists have long been protesting the historical exclusion and marginalization of people of color, women, minority groups, lower income groups, immigrants and other oppressed people in processes of urban development, whether in the historical institutional neglect of places and communities, or as part of focused urban sustainability or green planning interventions that affect them negatively (Agyeman & Evans, 2003; Park & Pellow, 2011; Pearsall & Anguelovski, 2016). Climate urbanism, or urban sustainability, as the "climate-inflected iteration of neoliberal urban development" (Robin & Castán Broto, 2021), is often seen to reproduce such inequalities, as in the example of eco- or green gentrification. This has been described as the process through which land, real estate and urban nature get reconfigured to make certain neighborhoods greener, climate-proof and thus more attractive while pushing out the most vulnerable longer-term residents. By stripping communities of their homes, livelihoods, and social and economic networks, these processes create "green" neighborhoods that are dominated by a white and rich "sustainability class" (Anguelovski et al., 2018; Gould & Lewis, 2016).

Building on these criticisms and adopting an analytical view that is both anti-colonial and feminist, political ecologists have counterposed an approach to adaptation that pays attention to structural and systemic preexisting inequalities by emphasizing the question, "Adaptation for whom?" (Cote & Nightingale, 2012; Nightingale, 2014). As they advocate, adaptation studies should approach climate impacts not as future and external threats, but as part of the embedded processes of racial capitalism, colonialism and other oppressions that have historically shaped today's unequal socio-environmental vulnerabilities (Ranganathan & Bratman, 2021). Adaptation should, thus, come "from below," departing from and integrating the existing knowledges and practices of local communities (Thorn, 2019).

In the urban realm, articulations of "transformative adaptation" propose drawing upon local experience and community knowledge, addressing underlying inequalities, and aiming at the creation and proliferation of new/alternative institutions, economies and behaviors (Chu et al., 2019). Some stress how a decolonizing and emancipatory approach to the urban planning example of greening cities in the US would mean not just that urban spaces become more inclusive and diverse "interfaces" for different races, genders and social classes, for example, but spaces for sharing the often uncomfortable stories of past and ongoing oppression, and for opening up the opportunities for "positive relationships of care and recognition" (Anguelovski et al., 2020, 2021). Elevating the voices, experiences and demands of vulnerable residents, critical scholars call for urban environmental justice that is not just about broadening or "equalizing" the distribution of environmental goods, including things such as climate-adapted infrastructure, nature-based solutions and greening, but for more deeply participatory, emancipatory, insubordinate, intersectional and relational approaches (ibid.).

Based on and synthesizing these advances, I argue, there cannot be a transformative vision of urban climate change adaptation that does not confront Europe's colonial legacies and ongoing stance toward the non-European "others" as they are continuously portrayed in European institutions and societies. In the next section, I sketch an approach to doing so by focusing on immigrant communities in European cities. I conceptualize immigrant communities as diverse kinds of collectivities that are not only highly vulnerable, but also, at the same time, social agents of change and equal knowledge-holders.

Postcolonial approaches to urban planning, sustainability, climate adaptation and justice, have so far focused on deconstructing and subverting the dominant narrative of "city-making" that focuses on the stories and trajectories of Western, so-called "global" cities. On the one hand, this includes an engagement with theory that goes beyond the traditional pillars of urban political ecology theory from and of cities in the Global North, thus a provincialization and situated-ness of urban theory (Lawhon et al., 2014). Such an approach would pay much more attention to the everyday practices, experiences and situated knowledge(s) circulating in the city, conceptualizing power in ways much more diffuse than concentrated, and understanding people "as the central means through which materials flow in many cities" (ibid.). It would allow for a different viewing of processes of urbanization in what has been called "ordinary cities" (Robinson, 2006), rather than "global cities," and seeing those ordinary places not as outliers or failures of the urban, but as equally valid, alternative and potentially subversive ways, trajectories and processes of city-making.

Similarly, when trying to define, capture and envision the materiality of urban sustainability, postcolonial approaches point to everyday attempts to manage, navigate, cope with and adapt to climate change that are emerging in places other than the Western global city and its mainstream models of climate adaptation. Such attempts and practices often escape the "spatial, temporal and subjective imaginations" (Kinnvall, 2016) that subconscious postcolonial legacies have imposed on the imaginaries of the urban and of climate urbanism. Robin and Castán Broto (2021) set out an agenda for a postcolonial analytical vocabulary that refocuses and reimagines urban responses to climate change beyond the North American and European model, and that engages more deeply with the ambiguities of climate urbanism. Moving away from colonial ideas where the Global North appears as having an exclusive and absolute monopoly over "civilization"—and thus also over "solutions" in the face of climate change—can reveal potentially transformative and more just responses that come from less visible places and communities. These responses might be less formal and spectacular, but also bear the potential of more freedom from capture by neoliberal dynamics.

Critical urban scholars thinking through postcolonial theory have suggested, rightly, that it is possible to inform the politics of urban climate adaptation from a different "place" and from a wider range of urban experiences. This possibility ought to be explored. Considering that environmental injustice (and environmental privilege) can operate on a variety of scales, both in and between the Global North and the Global South (Collins, 2010), I advance the argument that this possibility can emerge not only through a focus on the South, or on theories from the South, but also through a different epistemology that engages with the situated knowledge of migrant collectivities. That is, an approach to immigrants' knowledge (their embodied experiences) and immigrant knowledge (knowledge that travels), both of which move in and between the North and the South, the East and the West, the rural and the urban. I thus point to the ordinary and potentially subversive and transformational practices that can emerge *within* modern European cities, but from the position of the postcolonial subject of the immigrant. In fact, I argue, it is in this embodied experience of mobility, and of spatial and cultural crossings, that more inclusive and effective climate adaptation practices can emerge. First, by making visible the local material and embodied vulnerabilities produced by global political economies and ecologies. Second, and relatedly, by elevating the points of view and knowledge(s) of immigrant subjects concerning what constitutes vulnerability and how to address it, which have so far escaped the spotlight of the mainstream imaginaries of urban climate change adaptation.

In Europe during the last decade, amidst economic and health crises, and as right-wing political movements have expanded, immigration is increasingly being framed as a "crisis" and an economic, security and cultural threat (Tabaud, 2020). At the same time, discriminatory practices toward immigrants are intensifying along with deportations and detention centers where fundamental human rights and freedoms are restricted or denied (Open Access Now, 2014). Europeans tend to overestimate the number of immigrants in their countries, while even the net positive benefit that immigrants bring to national economies seems to be overlooked in mainstream discourse and public opinion (Bettini et al., 2021). As critical migration scholars have noted, it is not actual experiences, but "the national rule of territorialization ... and the resulting border and migration regimes which make migration and national border-crossing a problem" (Mecheril et al., 2013, cited in Gatt et al., 2016). Putting emphasis on immigration as a problem is more a political strategy than a collective misconception. Related to issues of public health, Kotsila and Kallis (2019) discuss how a politics of neoliberal austerity in Greece was as instrumental in creating the health crisis (including yearly outbreaks of malaria) as the politics of xenophobia and racism adopted by right-wing parties and governments. This intermingled and strategic political discourse exacerbated the health vulnerability of immigrants, and later stigmatized them as disease vectors, blaming them for their own vulnerability. Blaming immigration and immigrants themselves functions as a strategy for distracting attention not only from the real causes of suffered austerity, but also, as Malm et al. (2021) have argued, from the greater and encompassing crisis we are facing today, that of climate change. As such, it also distracts from the unevenly distributed impacts that climate change has on the health and well-being of workers, racialized groups, low-income and socioeconomically vulnerable people—categories that are often encapsulated in the figure of the non-EU immigrant. As recently noted by Bettini et al. (2021), in a context of multiple crises, it is this fear of "some othered migrant, of terrorism, of the erosion of livelihoods and welfare" that has been feeding the rise of populist right-wing parties and alt-right political movements in Europe, at the same time, when fear of climate change—although increasingly present—seems to be barely capable of mobilizing action at the level of local politics.

So far, the relationship between climate change and migration has focused mostly on places where migration *originates*, as places that have been rendered risky by climate change (among other factors). However true this might be, albeit mostly for semi-urban and urban migration (Tacoli, 2009), understanding this relationship only from this vantage point reinforces a view of migrants as passive victims of climatic, social and economic circumstances. It thus overlooks the fact that "migrants are also agents and subjects capable of acting" (Gatt et al., 2016, p. 3). In relation to climate change, recognizing immigration as a conscious and negotiated instantiation of agency can also open up our conceptualization of the relationship between climate change and immigrants in post-migration contexts. In other words, I argue, we need to start looking not only at how people migrate to escape climate change, but also how they adapt to it, take action for its mitigation, or mobilize for different climate politics, both before and *after* they migrate.

Immigrants are agents of ideas and practices that get exchanged and deployed between their new communities and those back home (Levitt & Lamba-Nieves, 2011). As such, they share knowledge that is both migrating and situated, both traveling and grounded. The relevance of such knowledge for addressing the climate change impacts and related health risks that are now facing European urban societies has so far received no attention in the literature (but see Klocker et al., 2018 for similar work in the rural agricultural sector). However, we know that when cities engage in dialogue with vulnerable communities, innovative ways for climate adaptation often emerge (UCCRN, 2018). The potential of immigrants as agents of positive change has mostly been studied in terms of the national economic benefits of immigration (Khojasteh & Raja, 2017). How often have we approached Bangladeshi or Moroccan communities in EU cities as experienced knowledge-holders regarding the best ways to deal with and adapt to long periods of extreme heat or severe flooding? Why haven't EU national and local governments consulted with people of Southeast Asian origin on their know-how regarding the best ways to manage the presence of mosquitoes as potential carriers of disease or the use of face masks as preventing airborne disease transmission? What are the behavioral strategies, infrastructural characteristics and mundane innovations that immigrants know due to previous exposure and cultural experience with different environments, or that they have developed because of being exposed to such risks both during their journeys and in the post-migration context? Asking these questions can reveal immigrant communities' rich knowledge, helping to develop and augment urban climate solutions that incorporate core concerns of immigrant groups around life and livelihood, while exposing and aiming to address structural the sociopolitical inequalities they experience. It can also reveal the structural drivers of unequal environmental and climate risks and the challenges standing in the way of immigrants implementing or accessing urban climate change adaptation.

Here I propose a postcolonial and feminist approach to embrace marginalized minorities' accounts of the problems of climate change and related health impacts, including their known ways of addressing them, the challenges involved and the changes they propose at the personal, collective and institutional levels. From a climate and health justice perspective, this would mean making visible and addressing immigrant communities' existing and exacerbated vulnerabilities, while also providing a counternarrative of agency that centers on immigrants' everyday lives, experiences and practices. Centering on the migrant and on knowledge informed by emigration and immigration, we move closer to what critical migration scholars have called a "migrant-centric epistemology" (Avallone & Gerbeau, 2021), one that is not constructed under the eyes of the West and that recognizes the heterogeneity of immigrant subjects.

The climate health vulnerabilities of immigrant populations through situated knowledge and everyday practices

How can we approach vulnerability to climate health risks through the ordinary everyday experience and lay knowledge of immigrant populations? When applying a cultural lens to the study of health, immigration scholars urge us to move beyond the framework of "acculturation" and instead bring together issues of structural racism, intersectionality and health vulnerability with immigrants' understandings and everyday practices (Verloo, 2013; Viruell-Fuentes et al., 2012). To begin with, this would require a recognition of historically produced climate, environmental and health injustices, as the long-lasting phenomena to be found in the everydayness of risk landscapes, constituted by "slow burning issues and slow disasters" (Privitera et al., 2021). Second, and relatedly, it would also mean seeing knowledge about climate health risk—vulnerability to it and adaptive capacity to resist it—as something performed in the everyday, as constructed and contested through subjective identities and relationships that hinge on power relations along lines of gender, class, nation and ethnicity, among others (Cote & Nightingale, 2012), and as memory and embodiment of risk. This would mean seeing knowledge not as a "thing," but as a process. Applying a cultural lens to questions of environmental and health risks is not new for critical medical anthropologists and, increasingly so, more common for feminist and health political ecologists. Researchers in these traditions have long proposed a focus beyond apolitical and universalizing accounts of disease or illness that, on the one hand, treat biomedicine as the only valid disciplinary standpoint and, on the other, tend to treat people as merely biophysical—and, in that respect, comparable—bodies, assigning little relevance to sociocultural context and difference. Instead, a focus on knowledge, discourse and identity is considered central in the understanding of health and disease dynamics in the context of socio-environmental change (Carroll, 2013; Connolly et al., 2017; Jackson & Neely, 2015; King, 2010; Lupton, 2015).

Donna Haraway (1988) writes: "There is no unmediated photograph or passive camera obscura in scientific accounts of bodies and machines; there are only highly specific visual possibilities, each with a wonderfully detailed, active, partial way of organizing worlds" (p. 583). In this sense, knowledge is necessarily partial and contextualized, whether it is about the riskiness of environments, disease cures, prevention methods or the physical impacts of illness. Building on Haraway's work, Nightingale (2016) posits that "in order to *see* adaptation, we need to probe discourses and perceptions of climate change, political economies, and social and political struggles, rather than biophysical change per se." Failing to look at climate health risk and the responses to it "from below," including through the experiences and value systems of marginalized and minority residents, precludes deeper understandings around health- and climate-related risks. Moreover, it constitutes a form of epistemic and hermeneutical injustice that can end up stigmatizing and dispossessing people of their agency to act in their best interest (Anguelovski et al., 2020; Glover, 2009).

A focus on situated knowledge as knowledge that reflects the particular perspectives of the knower and the context where it was created can allow for better interpretations of both climate health risk as a phenomenon and what might seem contradictory in terms of health behaviors or community responses to climate adaptation (Bee et al., 2015; Doshi, 2017; Rice et al., 2015). The credibility of information about health risk from the formal channels of public health institutions, for example, is closely related to trust in these institutions, as was shown in the case of water sanitation and diarrheal disease in poor rural communities in Vietnam (Kotsila, 2017) and heat-related preventive measures in multiethnic communities in Australia (Zografos et al., 2016). Investigating relationships of trust can reveal deeper issues of institutional racism, sociopolitical exclusion and environmental injustice that, in turn, help explain the health risk in question. Climate adaptation studies are increasingly recognizing the so-called intangible aspects of vulnerability such as risk perception, social capital and subjectivity formation (Adger, 2006; Cote & Nightingale, 2012; De Andrade & Szlafsztein, 2018), for being as important as—and related to—the more tangible aspects of vulnerability, such as poverty, living conditions, access to health-care services and chronic health issues. Interrogating both the tangible and intangible aspects in the case of climate-related health impacts is key when applying a situated knowledge approach to vulnerability. This would include asking how certain groups and individuals come to be positioned in relation to climate health adaptation efforts and to processes of capital accumulation in climate/health governance, and how they understand themselves within those processes since this may determine how they engage in or resist these efforts (Eriksen et al., 2015; Kaika, 2017).

Immigrants, in this case, may or may not be able to articulate the connection between an infectious vector-borne disease such as West Nile Virus (WNV) and climate change's effects on weather, in the same way this is done by academic institutions and documented in scientific reports. Their perceptions of the causes of WNV and its spread might be more raw, corporeal and direct: through personal experience and the meanings ascribed to it in public discourse, depending as well as on how this discourse operates in their community. Their embodied and situated knowledge of how to deal with diseases transmitted by mosquitoes might come from previous experience in places where such vectors thrive all year long, or from other types of diseases that have similar symptoms. It might also emerge indirectly from their accounts of everyday realities, such as their unsafe and precarious working conditions or limited health-care access options. These forms of knowledge hold important insights both for public health and for understanding how power works in relation to migration and intersectionality, through institutions and bodies, at various scales.

In the case of climate-related health vulnerabilities in urban environments, this points to an interdisciplinary research design that navigates public health data, health statistics, ethnographic observations, life stories and personal narrations of exposure to and experiences of risk, among other methods and approaches, "a kind of kaleidoscope wherein plural epistemologies help to reveal new, albeit partial and situated, patterns" (Nightingale, 2016). The everydayness of infection and sickness brings partial perspective into "objective" knowledge around disease and reveals how the personal is also political (Haraway, 1988). Such an approach also thus helps to locate individual bodies "within broader structures of power" relating the inequality of exposure "to conditions that produce poor health" (King, 2015), which encompass the social, environmental, cultural and political spheres, highlighting how vulnerability and adaptive capacity to climate change is a relation and not a condition or property of specific groups.

Conclusion

In this chapter, I have tried to draw attention to a blind spot in our understanding of the interface between climate change and migration, which I argue can be better explored if we amplify our points of view in epistemological terms through the approach of situated knowledge. Since climate impacts on human health are becoming more intense in cities globally, it is of utmost importance to generate knowledge that stems from the needs and interests of the most vulnerable communities. Immigrants, and especially immigrants from Majority World countries living in cities in Europe, are highly vulnerable groups due to structural inequalities that relate to citizenship rights, social exclusion, racial discrimination and deeper issues of unequal development, accumulation by dispossession and racism that characterize the global dynamics of capitalism. Invisibilizing immigrants in global urban climate adaptation and sustainability efforts is another way of reinforcing these dynamics of injustice, both by failing to understand how they are affected by climate change in post-migration contexts, and by failing to include their understandings, needs and behaviors in assessments of vulnerability and plans for adaptation.

In sum, this chapter is a call for research on climate health vulnerability and adaptation that takes an approach to immigrants as knowledgeable and equal agents of change. Such research would apply a feminist postcolonial lens to vulnerability and adaptation that focuses on the broader tangible and intangible vulnerabilities[1] that characterize urban climate health risks, and investigate deeply immigrant communities' embodied, experiential and practiced knowledge about adaptation to climate risks. Under this rubric, future research should focus on these questions: How vulnerable are immigrants to climate health risks in migration destinations, and what are the broader social, political and ecological drivers of this vulnerability? How is the adaptive capacity of immigrants shaped by the knowledge and practices they share and can or cannot employ in response to such risks? How does an approach to situated—that is, embodied, partial and accountable—knowledge(s) contribute to a politicized understanding of adaptation dynamics in relation to immigration?

References

Adger, W. N. (2006). Vulnerability. *Global Environmental Change, 16*(3), 268–281. https://doi.org/10.1016/j.gloenvcha.2006.02.006

Adger, W. N., Safra de Campos, R., & Mortreux, C. (2018). Mobility, displacement and migration, and their interactions with vulnerability and adaptation to environmental risks. In R. McLeman & F. Gemenne (Eds.) *The Routledge handbook of environmental displacement and migration* (pp. 29–41). Routledge.

Agyeman, J., & Evans, T. (2003). Toward just sustainability in urban communities: Building equity rights with sustainable solutions *Annals of the American Academy of Political and Social Science, 590*(1), 35–53. https://doi.org/10.1177/0002716203256565

Agyeman, J., Schlosberg, D., Craven, L., & Matthews, C. (2016). Trends and directions in environmental justice: From inequity to everyday life, community, and just

1 A primary set of vulnerability indicators that could apply to climate health risks from such an approach would, for example, include aspects of livelihood and occupation, neighborhood amenities and housing conditions, socioeconomic status, health status and health care, health risk perceptions, demographic characteristics, immigration characteristics, knowledge based on previous experiences, knowledge about prevention or treatments and access to social capital.

sustainabilities. *Annual Review of Environment and Resources, 41*, 321–340. https://doi.org/10.1146/annurev-environ-110615-090052

Andreucci, D., & Zografos, C. (2022). Between improvement and sacrifice: Othering and the (bio)political ecology of climate change. *Political Geography, 92*, 102512. https://doi.org/10.1016/j.polgeo.2021.102512

Anguelovski, I. (2016). Healthy food stores, greenlining and food gentrification: Contesting new forms of privilege, displacement and locally unwanted land uses in racially mixed neighborhoods. *International Journal of Urban and Regional Research, 39*(6), 1209–1230. https://doi.org/10.1111/1468-2427.12299

Anguelovski, I., Brand, A. L., Connolly, J. J. T., Corbera, E., Kotsila, P., Steil, J., Garcia-Lamarca, M., Triguero-Mas, M., Cole, H., & Baró, F. (2020). Expanding the boundaries of justice in urban greening scholarship: Toward an emancipatory, antisubordination, intersectional, and relational approach. *Annals of the American Association of Geographers, 110*(6), 1743–1769. https://doi.org/10.1080/24694452.2020.1740579

Anguelovski, I., Brand, A. L., Ranganathan, M., & Hyra, D. (2021). Decolonizing the green city: From environmental privilege to emancipatory green justice. *Environmental Justice, 15*(1), 1–11. https://doi.org/10.1089/env.2021.0014

Anguelovski, I., Connolly, J., & Brand, A. L. (2018). From landscapes of utopia to the margins of the green urban life. *City, 22*(3), 417–436. https://doi.org/10.1080/13604813.2018.1473126

Anguelovski, I., Connolly, J. J. T., Pearsall, H., Shokry, G., Checker, M., Maantay, J., Gould, K., Lewis, T., Maroko, A., & Roberts, J. T. (2019). Opinion: Why green "climate gentrification" threatens poor and vulnerable populations. *Proceedings of the National Academy of Sciences, 116*(52), 26139–26143. https://doi.org/10.1073/pnas.1920490117

Åström, C., Orru, H., Rocklöv, J., Strandberg, G., Ebi, K. L., & Forsberg, B. (2013). Heat-related respiratory hospital admissions in Europe in a changing climate: A health impact assessment. *BMJ Open, 3*, e001842. https://doi.org/10.1136/bmjopen-2012-0018423(1)

Avallone, G., & Gerbeau, Y. M. (2021). Liberar las migraciones: la contribución de Abdelmalek Sayad a una epistemología migrante-céntrica. *Migraciones Internacionales, 12*. https://doi.org/10.33679/rmi.v1i1.1949

Bakonyi, T., & Haussig, J. M. (2020). West Nile virus keeps on moving up in Europe. *Euro Surveillance: Bulletin europeen sur les maladies transmissibles/European Communicable Disease Bulletin, 25*(46), 2001938. https://doi.org/10.2807/1560-7917.ES.2020.25.46.2001938

Bee, B. A., Rice, J., & Trauger, A. (2015). A feminist approach to climate change governance: Everyday and intimate politics. *Geography Compass, 9*(6), 339–350. https://doi.org/10.1111/gec3.12218

Benach, J., Muntaner, C., Chung, H., & Benavides, F. G. (2010). Immigration, employment relations, and health: Developing a research agenda. *American Journal of Industrial Medicine, 53*(4), 338–343. https://doi.org/10.1002/ajim.20717

Bettini, G., Beuret, N., & Turhan, E. (2021). On the frontlines of fear: Migration and climate change in the local context of Sardinia, Italy. *ACME: An International Journal for Critical Geographies, 20*(3 SE-Research). https://acme-journal.org/index.php/acme/article/view/1838

Bulkeley, H., Edwards, G. A. S., & Fuller, S. (2014). Contesting climate justice in the city: Examining politics and practice in urban climate change experiments. *Global Environmental Change, 25*, 31–40. https://doi.org/10.1016/j.gloenvcha.2014.01.009

Bullard, R. D., & Lewis, J. (1996). *Environmental justice and communities of color.* Sierra Club Books.

Carroll, J. J. (2013). Key theories from critical medical anthropology for public health research, Part I: Starting with Foucault: Cultures of medicine and meanings of illness. *Tobacco Control and Public Health in Eastern Europe, 3*(1), 39–46.

Castañeda, H., Holmes, S. M., Madrigal, D. S., Young, M.-E. D., Beyeler, N., & Quesada, J. (2015). Immigration as a social determinant of health. *Annual Review of Public Health, 36*, 375–392. https://doi.org/10.1146/annurev-publhealth-032013-182419

Chu, E., Brown, A., Michael, K., Du, J., Lwasa, S., & Mahendra, A. (2019). Unlocking the potential for transformative climate adaptation in cities. Background paper prepared for the Global Commission on Adaptation, World Resources Institute, Washington, DC, and Rotterdam. https://wrirosscities.org/sites/default/files/FINAL19_GCA_Cities_Background%20Paper.pdf

Chun, J. M. (2015, June). *Planned relocations in the Mekong Delta, Vietnam: A successful model for climate change adaptation, a cautionary tale, or both?* Brookings–LSE Project on Internal Displacement. https://www.brookings.edu/research/planned-relocations-in-the-mekong-delta-vietnam-a-successful-model-for-climate-change-adaptation-a-cautionary-tale-or-both/

Collins, T. W. (2010). Marginalization, facilitation, and the production of unequal risk: The 2006 Paso del Norte Floods. *Antipode, 42*(2), 258–288. https://doi.org/10.1111/j.1467-8330.2009.00755.x

Connolly, C., Kotsila, P., & D'Alisa, G. (2017). Tracing narratives and perceptions in the political ecologies of health and disease. *Journal of Political Ecology, 24*(1), 1–10. https://doi.org/10.2458/v24i1.20778

Connolly, J. J. T. (2018). From Jacobs to the just city: A foundation for challenging the green planning orthodoxy. *Cities, 91*, 64–70. https://doi.org/10.1016/j.cities.2018.05.011

Cote, M., & Nightingale, A. J. (2012). Resilience thinking meets social theory: Situating social change in socio-ecological systems (SES) research. *Progress in Human Geography, 36*(4), 475–489. https://doi.org/10.1177/0309132511425708

De Andrade, M. M. N., & Szlafsztein, C. F. (2018). Vulnerability assessment including tangible and intangible components in the index composition: An Amazon case study of flooding and flash flooding. *Science of the Total Environment, 630*, 903–912. https://doi.org/10.1016/j.scitotenv.2018.02.271

Doshi, S. (2017). Embodied urban political ecology: Five propositions. *Area, 49*(1), 125–128. https://doi.org/10.1111/area.12293

EC. (2019). *The European Green Deal*. European Commission. https://commission.europa.eu/strategy-and-policy/priorities-2019-2024/european-green-deal_en

EC. (2020). Adaptation to health effects of climate change in Europe. Group of Chief Scientific Advisors, European Commission. https://data.europa.eu/doi/10.2777/869383

Eriksen, S. H., Nightingale, A. J., & Eakin, H. (2015). Reframing adaptation: The political nature of climate change adaptation. *Global Environmental Change, 35*, 523–533. https://doi.org/10.1016/j.gloenvcha.2015.09.014

Gamble, J. L., Balbus, J., Berger, M., Bouye, K., Campbell, V., Chief, K., Conlon, K., Crimmins, A., Flanagan, B., & Gonzalez-Maddux, C. (2016). *The impacts of climate change on human health in the United States: A scientific assessment*. US Global Change Research Program. https://health2016.globalchange.gov/low/ClimateHealth2016_FullReport_small.pdf

Gatt, S., Hazibar, K., Sauermann, V., Preglau, M., & Ralser, M. (2016). Migration from a gender-critical, postcolonial and interdisciplinary perspective. *Österreichische Zeitschrift Für Soziologie, 41*(3), 1–12. https://doi.org/10.1007/s11614-016-0236-4

Glover, S. M. (2009). Review of Mark Nichter, *Global health: Why cultural perceptions, social representations, and biopolitics matter. Human Ecology, 37*(5), 669–670. https://doi.org/10.1007/s10745-009-9242-5

Gould, K. A., & Lewis, T. L. (2016). *Green gentrification: Urban sustainability and the struggle for environmental justice*. Routledge.

Haddad, M. (2021, July 1). Mapping the hottest temperatures around the world. *Al Jazeera*. https://www.aljazeera.com/news/2021/7/1/interactive-mapping-hottest-temperatures-around-world

Hansen, A., Bi, L., Saniotis, A., & Nitschke, M. (2013). Vulnerability to extreme heat and climate change: Is ethnicity a factor? *Global Health Action, 6*(1), 21364. https://doi.org/10.3402/gha.v6i0.21364

Haraway, D. (1988). Situated knowledges: The science question in feminism and the privilege of partial perspective. *Feminist Studies, 14*(3), 575–599. https://doi.org/10.2307/3178066

Hemminki, K. (2014). Immigrant health, our health. *European Journal of Public Health, 24*(suppl_1), 92–95. https://doi.org/10.1093/eurpub/cku108

Heynen, N., Kaika, M., & Swyngedouw, E. (Eds.). (2006). *In the nature of cities: Urban political ecology and the politics of urban metabolism*. Routledge.

IOM. (2015). *IOM outlook on migration, environment and climate change.* International Organisation for Migration.

Jackson, P., & Neely, A. H. (2015). Triangulating health toward a practice of a political ecology of health. *Progress in Human Geography, 39*(1), 47–64. https://doi.org/10.1177/0309132513518832

Kaika, M. (2017). "Don't call me resilient again!": The New Urban Agenda as immunology ... or ... what happens when communities refuse to be vaccinated with "smart cities" and indicators. *Environment and Urbanization, 29*(1), 89–102. https://doi.org/10.1177/0956247816684763

Khojasteh, M., & Raja, S. (2017). Agents of change: How immigrant-run ethnic food retailers improve food environments. *Journal of Hunger & Environmental Nutrition, 12*(3), 299–327. https://doi.org/10.1080/19320248.2015.1112759

King, B. (2010). Political ecologies of health. *Progress in Human Geography, 34*(1), 38–55. https://doi.org/10.1177/0309132509338642

King, B. (2015). Political ecologies of disease and health. In T. Perreault, G. Bridge, & J. McCarthy (Eds.), *The Routledge handbook of political ecology* (pp. 343–353). Routledge.

Kinnvall, C. (2016). The postcolonial has moved into Europe: Bordering, security and ethno-cultural belonging. *JCMS: Journal of Common Market Studies, 54*(1), 152–168. https://doi.org/10.1111/jcms.12326

Klocker, N., Head, L., Dun, O., & Spaven, T. (2018). experimenting with agricultural diversity: Migrant knowledge as a resource for climate change adaptation. *Journal of Rural Studies, 57*, 13–24. https://doi.org/10.1016/j.jrurstud.2017.10.006

Kotsila, P. (2017). Health dispossessions and the moralization of disease: The case of diarrhea in the Mekong Delta, Vietnam. *Journal of Political Ecology, 24*(1), 97–109. https://doi.org/I:10.2458/v24i1.20785

Kotsila, P., & Kallis, G. (2019). Biopolitics of public health and immigration in times of crisis: The malaria epidemic in Greece (2009–2014), *Geoforum, 106*, 223–233. https://doi.org/10.1016/j.geoforum.2019.08.019

Lassetter, J. H., & Callister, L. C. (2008). The impact of migration on the health of voluntary migrants in Western societies: A review of the literature. *Journal of Transcultural Nursing, 20*(1), 93–104. https://doi.org/10.1177/1043659608325841

Lawhon, M., Ernstson, H., & Silver, J. (2014). Provincializing urban political ecology: Towards a situated UPE through African urbanism. *Antipode, 46*(2), 497–516. https://doi.org/10.1111/anti.12051

Lay, C. R., Sarofim, M. C., Zilberg, A. V., Mills, D. M., Jones, R. W., Schwartz, J., & Kinney, P. L. (2021). City-level vulnerability to temperature-related mortality in the USA and future projections: A geographically clustered meta-regression. *The Lancet Planetary Health, 5*(6), e338-e346. https://doi.org/10.1016/S2542-5196(21)00058-9

Lebano, A., Hamed, S., Bradby, H., Gil-Salmerón, A., Durá-Ferrandis, E., Garcés-Ferrer, J., Azzedine, F., Riza, E., Karnaki, P., Zota, D., & Linos, A. (2020). Migrants' and refugees' health status and healthcare in Europe: A scoping literature review. *BMC Public Health, 20*(1), 1039. https://doi.org/10.1186/s12889-020-08749-8

Levitt, P., & Lamba-Nieves, D. (2011). Social remittances revisited. *Journal of Ethnic and Migration Studies, 37*(1), 1–22. https://doi.org/10.1080/1369183X.2011.521361

Lupton, D. (2015). The pedagogy of disgust: The ethical, moral and political implications of using disgust in public health campaigns. *Critical Public Health, 25*(1), 4–14. https://doi.org/10.1080/09581596.2014.885115

Macintyre, H. L., Heaviside, C., Taylor, J., Picetti, R., Symonds, P., Cai, X.-M., & Vardoulakis, S. (2018). Assessing urban population vulnerability and environmental risks across an urban area during heatwaves: Implications for health protection. *Science of The Total Environment, 610-611*, 678–690. https://doi.org/10.1016/j.scitotenv.2017.08.062

Malm, A., and the Zetkin Collective. (2021). *White skin, black fuel: On the danger of fossil fascism*. Verso.

Marinucci, G. D., Luber, G., Uejio, C. K., Saha, S., & Hess, J. J. (2014). Building resilience against climate effects: A novel framework to facilitate climate readiness in public health agencies. *International Journal of Environmental Research and Public Health, 11*(6), 6433–6458. https://doi.org/10.3390/ijerph110606433

MedECC. (2020). *Climate and Environmental Change in the Mediterranean Basin: Current Situation and Risks for the Future. First Mediterranean Assessment Report* [Cramer, W., Guiot, J., & Marini, K. (Eds.)]. Union for the Mediterranean, Plan Bleu, UNEP/MAP. https://doi.org/10.5281/zenodo.4768833

Millington, N. (2018). Linear parks and the political ecologies of permeability: Environmental displacement in São Paulo, Brazil. *International Journal of Urban and Regional Research, 42*(5), 864–881. https://doi.org/10.1111/1468-2427.12657

Mills-Novoa, M., Boelens, R., Hoogesteger, J., & Vos, J. (2020). Governmentalities, hydrosocial territories & recognition politics: The making of objects and subjects for climate change adaptation in Ecuador. *Geoforum, 115*, 90–101. https://doi.org/10.1016/j.geoforum.2020.06.024

Nightingale, A. J. (2014). A socionature approach to adaptation. In T. Håkon Inderberg, S. Eriksen, K. O'Brien, & L. Sygna (Eds.), *Climate change adaptation and development: Transforming paradigms and practices* (pp. 219–234). Routledge.

Nightingale, A. J. (2016). Adaptive scholarship and situated knowledges? Hybrid methodologies and plural epistemologies in climate change adaptation research. *Area, 48*(1), 41–47. https://doi.org/10.1111/area.12195

Open Access Now. (2014). *The hidden face of immigration detention camps in Europe*. http://www.migreurop.org/IMG/pdf/hiddenfaceimmigrationcamps-okweb.pdf

Park, L. S.-H., & Pellow, D. N. (2011). *The slums of Aspen: Immigrants vs. the environment in America's Eden*. New York University Press.

Paz, S. (2019). Effects of climate change on vector-borne diseases: An updated focus on West Nile virus in Humans. *Emerging Topics in Life Sciences, 3*(2), 143–152. https://doi.org/0.1042/ETLS20180124

Pearsall, H., & Anguelovski, I. (2016). Contesting and resisting environmental gentrification: Responses to new paradoxes and challenges for urban environmental justice. *Sociological Research Online, 21*(3), 1–7. https://doi.org/10.5153/sro.3979

Pearsall, H., & Pierce, J. (2010). Urban sustainability and environmental justice: Evaluating the linkages in public planning/policy discourse. *Local Environment, 15*(6), 569–580. https://doi.org/10.1080/13549839.2010.487528

Privitera, E., Armiero, M., & Gravagno, F. (2021). Seeking justice in risk landscapes: Small data and toxic autobiographies from an Italian petrochemical town (Gela, Sicily). *Local Environment, 26*(7), 847–871. https://doi.org/0.1080/13549839.2021.1922999

Pulido, L. (2010). Rethinking environmental racism: White privilege and urban development in southern California. *Annals of the Association of American Geographers, 90*(1), 12–40. https://doi.org/10.1111/0004-5608.00182

Ramamurthy, P., & Bou-Zeid, E. (2017). Heatwaves and urban heat islands: A comparative analysis of multiple Cities. *Journal of Geophysical Research: Atmospheres, 122*(1), 168–178. https://doi.org/10.1002/2016JD025357

Ramirez, M. D., Liebig, T., Thoreau, C., & Veneri, P. (2018). The integration of migrants in OECD regions: A first assessment. OECD Regional Development Working Papers 1. https://www.oecd-ilibrary.org/urban-rural-and-regional-development/the-integration-of-migrants-in-oecd-regions_fb089d9a-en

Ranganathan, M., & Bratman, E. (2021). From urban resilience to abolitionist climate justice in Washington, DC. *Antipode, 53*(1), 115–137. https://doi.org/10.1111/anti.12555

Rice, J. L., Burke, B. J., & Heynen, N. (2015). Knowing climate change, embodying climate praxis: Experiential knowledge in southern Appalachia, *Annals of the Association of American Geographers, 105*(2), 253–262. https://doi.org/10.1080/00045608.2014.985628

Robin, E., & Castán Broto, V. (2021). Towards a postcolonial perspective on climate urbanism. *International Journal of Urban and Regional Research, 45*(5), 869–878. https://doi.org/10.1111/1468-2427.12981

Robinson, J. (2006). *Ordinary cities: Between modernity and development*. Routledge.

Rocheleau, D. (2015). A situated view of feminist political ecology from my networks, roots and territories. In W. Harcourt and I. L. Nelson (Eds.), *Practising feminist political ecologies: Moving beyond the "green economy"* (pp. 29–66). Zed Books.

Rocque, R. J., Beaudoin, C., Ndjaboue, R., Cameron, L., Poirier-Bergeron, L., Poulin-Rheault, R.-A., Fallon, C., Tricco, A. C., & Witteman, H. O. (2021). Health effects of climate change: An overview of systematic reviews, *BMJ Open, 11*(6), e046333. https://doi.org/10.1136/bmjopen-2020-046333

Shi, L., Chu, E., Anguelovski, I., Aylett, A., Debats, J., Goh, K., Schenk, T., Seto, K. C., Dodman, D., Roberts, D., Roberts, J. T., & VanDeveer, S. D. (2016). Roadmap towards justice in urban climate adaptation research. *Nature Climate Change, 6*(2), 131–137. https://doi.org/10.1038/nclimate2841

Shokry, G., Connolly, J. J., & Anguelovski, I. (2020). Understanding climate gentrification and shifting landscapes of protection and vulnerability in green resilient Philadelphia. *Urban Climate, 31*, 100539. https://doi.org/10.1016/j.uclim.2019.100539

Silver, J. (2019). Decaying infrastructures in the post-industrial city: An urban political ecology of the US pipeline crisis. *Environment and Planning E: Nature and Space, 4*(3), 756–777. https://doi.org/10.1177/2514848619890513

Smith, K., Woodward, A., Campbell-Lendrum, D., Chadee, D., Honda, Y., Liu, Q., Olwoch, J., Revich, B., Sauerborn, R., & Aranda, C. (2014). Human health: Impacts, adaptation, and co-benefits. In *Climate change 2014: Impacts, adaptation, and vulnerability. Part A: Global and sectoral aspects. Contribution of Working Group II to the Fifth Assessment Report of the Intergovernmental Panel on Climate Change* (pp. 709–754). Cambridge University Press.

Stripple, J., & Bulkeley, H. (2015). Governmentality. In K. Bäckstrand & E. Lövbrand (Eds.), *Research handbook on climate governance* (pp. 49–59). Edward Elgar.

Tabaud, A. L. (2020). *Explaining the main drivers of anti-immigration attitudes in Europe.* Eyes on Europe. https://www.eyes-on-europe.eu/explaining-the-main-drivers-of-anti-immigration-attitudes-in-europe/

Tacoli, C. (2009). Crisis or adaptation? Migration and climate change in a context of high mobility. *Environment and Urbanization, 21*(2), 513–525. https://doi.org/10.1177/0956247809342182

Taylor, M. (2014). *The political ecology of climate change adaptation: Livelihoods, agrarian change and the conflicts of development.* Routledge.

Thorn, J. P. R. (2019). Adaptation "from below" to changes in species distribution, habitat and climate in agro-ecosystems in the Terai plains of Nepal. *Ambio, 48*(12), 1482–1497. https://doi.org/10.1007/s13280-019-01202-0

UCCRN. (2018). *IMPACT 2050: The future of cities under climate change.* https://c40-production-images.s3.us-west-2.amazonaws.com/impact2050report.pdf

Verloo, M. (2013). Intersectional and cross-movement politics and policies: Reflections on current practices and debates. *Signs: Journal of Women in Culture and Society, 38*(4), 893–915. https://doi.org/10.1086/669572

Viruell-Fuentes, E. A., Miranda, P. Y., & Abdulrahim, S. (2012). More than culture: Structural racism, intersectionality theory, and immigrant health. *Social Science & Medicine, 75*(12), 2099–2106. https://doi.org/10.1016/j.socscimed.2011.12.037

Wacquant, L., Slater, T., & Pereira, V. B. (2014). Territorial stigmatization in action. *Environment and Planning A, 46*(6), 1270–1280. https://doi.org/10.1068/a4606ge

Watts, M. J. (2015). Now and then: The origins of political ecology and the rebirth of adaptation as a form of thought. In T. Perreault, G. Bridge, & J. McCarthy (Eds.), *The Routledge handbook of political ecology* (pp. 19–50). Routledge.

Watts, M. J., Kotsila, P., Mortyn, P. G., Sarto i Monteys, V., & Urzi Brancati, C. (2020a). Influence of socio-economic, demographic and climate factors on the regional distribution of dengue in the United States and Mexico. *International Journal of Health Geographics, 19*(44). https://doi.org/10.1186/s12942-020-00241-1

Watts, N., Amann, M., Arnell, N., Ayeb-Karlsson, S., Beagley, J., Belesova, K., Boykoff, M., Byass, P., Cai, W., & Campbell-Lendrum, D. (2020b). The 2020 report of *The Lancet* countdown on health and climate change: Responding to converging crises. *The Lancet, 9*(397),129–170. https://doi.org/10.1016/S0140-6736(20)32290-X

WHO. (2010). *Parma Declaration on Environment and Health, Parma, Italy, 10–12 March 2010*. WHO Regional Office for Europe.

WHO. (2017). *Protecting health in Europe from climate change: 2017 update*. https://apps.who.int/iris/handle/10665/329522

WHO. (2018). *Public health and climate change adaptation policies* (*Final report*). WHO Press.

Woodruff, S. C., Meerow, S., Stults, M., & Wilkins, C. (2018). Adaptation to resilience planning: Alternative pathways to prepare for climate change. *Journal of Planning Education and Research, 42*(1), 64–75. https://doi.org/10.1177/0739456X18801057

Zografos, C., Anguelovski, I., & Grigorova, M. (2016). When exposure to climate change is not enough: Exploring heatwave adaptive capacity of a multi-ethnic, low-income urban community in Australia. *Urban Climate, 17*, 248–265. https://doi.org/10.1016/j.uclim.2016.06.003

About the Author

Panagiota Kotsila is a researcher at the Barcelona Laboratory for Urban Environmental Justice and Sustainability (BCNUEJ) at the Institute of Environmental Science and Technology (ICTA-UAB), Autonomous University of Barcelona, Spain. She researches political ecologies of health and urban environmental justice. She recently coauthored the book *Injustice in Urban Sustainability* (with I. Anguelovski, M. Garcia-Lamarca and F. Sekulova, Routledge, 2022).

10. Small towns facing big problems

Sustainable development, social choice and the challenge of local-level organizing for the environment: Insights from Flagler Beach, Florida, USA

Chad Boda

Abstract
Managing tensions between the need for economic development and urgently necessary environmental protection is a long-standing challenge for policymaking. By drawing on the case of coastal erosion in Flagler Beach based on five years of research, Boda argues that social choice, which is a decision-making procedure developed within the capabilities approach to human development, is a good fit to solve this tension. Social movements are of particular importance here since they can expand and build new alliances and even explore more radical forms of knowledge coproduction. By identifying contextually relevant decision criteria through reasoned public deliberation, social choice can help policymakers, civil society and all relevant stakeholders to open space for deliberation and experimentation.

Keywords: sustainable development, coastal erosion, collective action, social choice, Florida

It hardly needs mentioning that our time is one characterized by crises of global environmental change and gross political-economic disparities. These intertwined problems are ultimately challenges of sustainable development. Sustainable development is here understood as a contested concept, though with common roots in the Brundtland Commission and its recognition of the challenge of meeting human needs now and in the future. Ultimately, this is a challenge of managing tensions between the need for economic development and increasingly urgent environmental conservation. The

---

Armiero, M., E. Turhan & S. Paolo De Rosa (eds), *Urban Movements and Climate Change: Loss, Damage and Radical Adaptation*. Taylor & Francis Group 2023
DOI: 10.5117/9789463726665_CH10

question that any theory of sustainable development should answer is "What should be sustained?" when faced with such trade-offs, and from this follow appropriate metrics and tools for decision-making. A variety of competing approaches have been proposed (see Boda & Faran, 2018, Part II). These can be divided into two main approaches: capital theory, the most dominant approach both in theory and practice, and the capabilities approach to human development, a marginal, but promising alternative applied theoretically in some areas such as disaster risk reduction (see Gardoni & Murphy, 2008, 2009, 2010), although employed less frequently in practice (Boda et al., 2021). An approach to sustainable development is determined by its concepts, metrics and policies (Boda et al., 2020). Different approaches to sustainable development can lead to incompatible or contradictory accounts of the nature of the problem and what should be done about it, and thus to political conflict over alternatives.

In this chapter, I discuss these challenges in relation to a case of critical erosion management in the coastal city of Flagler Beach, located in Flagler County, northeastern Florida, USA. Drawing on five years of research (see Boda, 2018a), public opinion surveys and social movement theory in the political process tradition, this chapter has two primary aims: (1) to argue for the relevance of social choice as a decision-making procedure for communities to address sustainable development problems like climate change adaptation, and (2) to point out some of the challenges faced local civil society organizations are likely to face if they hope to make social choice a reality.

Flagler Beach, critical erosion and the necessity of social choice

Flagler Beach is a small beachside town with a permanent population of less than 5,000 residents. It is located in Flagler County, which has been one of the fastest growing areas in the state for the last three decades. The main source of revenue for county and local governments in Florida comes from property and sales tax, which puts serious constraints on local-level capacity to generate revenue, forcing many to prioritize tourism sales tax as a main source of revenue generation and expansion. The main draw for tourists to the region are its beaches, which sport a unique form of red shell beach sand known as coquina, and have long been a popular regional swimming, fishing and surfing destination. An increasing focus on tourism has characterized Flagler County for some time, particularly since the 2008 financial crisis when Flagler County was among the hardest hit counties in the country in terms of revenue loss and unemployment (see Boda, 2018b).

This pressure to increase tourism was felt countywide, particularly in Flagler Beach, where the lack of alternative industries means tourism is the de facto lifeline for the local economy. Within this context, Flagler Beach has long battled beach erosion, which over the years has threatened, among other things, State Road A1A, the coastal byway that is the crucial access route for tourists as well as a hurricane evacuation route and a state-recognized historic scenic byway.

The main cause of erosion in the city is a combination of the placement of State Road A1A, which is located directly on top of the barrier island foredune, and natural erosional processes, where erosion control measures have led to the beach being starved of the sediment it needs to regenerate after normal erosion events (see Boda, 2018a, Section 4.2). However, even though the placement of State Road A1A is acknowledged by local, state and federal authorities to be a major driver of beach erosion, its economic and cultural importance mean that moving the road is an extremely unpopular idea. As a result, the erosion problem has only gotten worse over the last half-century, particularly given the effects of sporadic hurricanes (such as those of the 2004–2005 season and Hurricane Matthew in 2016), which are likely to increase in severity due to climate change, adding to the urgency of coastal adaptation. In its attempts to maintain the structural integrity of State Road A1A, the Florida Department of Transportation (FDOT), which maintains primary responsibility for the road, has historically used rock revetments and small seawall segments to keep the road from collapsing, which has led to large portions of the city's coastline to be designated as "critically eroded," threatening recreational beach and nesting habitat for endangered sea turtle species in addition to private and public infrastructure. This strategy has been extremely controversial over the years (Boda, 2015). As an alternative to the continued maintenance of rock revetments, after the 2004 hurricane season, the US Army Corps of Engineers (USACE) initiated a 10-year feasibility study to consider the viability of a large-scale beach renourishment project, which has never come to fruition, but is still in the cards. The possibility of large-scale beach renourishment has also been a major point of controversy in the town (Boda, 2018c).

The approach to sustainable development, whether conscious or implicit, is not separate from or complementary to climate change mitigation, adaptation or loss and damage, it *encompasses* all of them (Boda et al., 2020). The answer to "What should be sustained?" defines how one should decide between trade-offs in mitigation and adaptation, or how to account for and address loss and damage. In Flagler Beach, the dominant approach has been capital theory, which aims to sustain human welfare by sustaining economic productivity, and this has translated into the FDOT seawall and the USACE renourishment project, both of which have been justified by cost-benefit analyses. While both of these approaches have been justified economically, they lead to residual environmental and social concerns. Other citizen concerns, including environmental conservation, recreation and place-attachment value, are subordinate to the maximization of benefits over costs, which is a legal requirement for these agencies (Boda, 2018c). The neglect of social and environmental concerns follows directly from the economic choice approach to collective decision-making and the tool of cost-benefit analysis. Overcoming these limitations will require using a different set of decision-making criteria in an enabling institutional environment.

Sustainable development and collective decision-making

In the United States, environmental policy assessment has long been dominated by cost-benefit analysis, rather than, for example, the social acceptability of outcomes, ecological units or other nonutilitarian criteria. This prioritization of economic reasoning has provided the justification for environmentally and socially problematic solutions such as the FDOT seawall projects mentioned above. The economic choice approach to environmental policy evaluation, i.e., capital theory, has well-known and major limitations. These shortcomings are both general (Boda et al., 2021, 2022b) and specific to coastal management in Florida (Boda, 2018b, 2018c), and include the propensity to exclude noneconomic concerns, producing inaccurate social valuations of nonmarket goods and services and neglecting the issue of the distribution of costs and benefits within society (Boda, 2018c).

In light of these criticisms, the capabilities approach has been developed as an alternative to the economic reasoning of capital theory. This approach is based on a recognition that there is a wide span of relevant information beyond monetary value that should be considered in collective decision-making, including, for example, peoples' health, environmental quality, political freedoms, etc. This in turn requires that society's wide variety of valid concerns be evaluated and prioritized through public deliberation. Rather than relying on a unitary metric such as monetary value as in capital theory, this approach applies the unifying concept of capabilities (see Sen, 2001), which coherently expresses the relations between heterogenous life conditions and values without reducing them to the same metric. When making collective decisions in this tradition, there will always be trade-offs between the capabilities of different segments in a community, both within and between generations, which are decided on the basis of social reasoning and acceptability, rather than predefined formulas such as cost-benefit analysis).

Amartya Sen has elaborated the concept of "social choice" (Sen, 1999) as an appropriate strategy for operationalizing the capabilities approach. In effect, social choice is a collective decision-making procedure through which socially acceptable criteria can be identified and applied to the evaluation of alternative projects or development pathways. As Sen has put it,

> It is crucial to ask, in any evaluative exercise, how the weights are to be selected. This judgmental exercise can be resolved only through reasoned evaluation. For a particular person, who is making his or her own judgments, the selection of weights will require reflection, rather than any interpersonal agreement (or consensus). However, in arriving at an "agreed" range for social evaluation, there has to be some kind of a reasoned "consensus" on weights, or at least on a range of weights. This is a "social choice" exercise, and it requires public discussion and a democratic understanding and acceptance. (Sen, 2011, pp. 78–79)

There are many good reasons to take social choice seriously as a viable and superior alternative to the universal application of cost-benefit analysis. But how can this promising alternative come about? Clearly, getting to social choice requires social change, which itself requires social action.

Implementing social choice: The role of civil society

The operationalization of social choice in pursuit of sustainable development would depend on the presence of three basic requirements (see Boda, 2018c; also see Sen, 2001, p. 249), namely:

1. The production of an adequate evaluative framework: as in any decision-making procedure, social choice requires the application of an evaluative framework to decide between competing alternatives. However, rather than prescribing a predefined formula like cost-benefit analysis, this framework requires that citizens participate in identifying the various evaluative criteria to be applied when judging between given alternatives, so that these criteria and their respective weights reflect contextually relevant capabilities, functioning states and value sets. The

process of identifying criteria and weights is necessarily one of open public deliberation, where citizens express and debate not only what their preferences are, but the values and reasons that inform them (see Sen, 1999).

2. The adherence of participating individuals and agencies to conducive norms of behavior and reasoning, the production of such an evaluative framework further requires that individuals engage in evaluative processes as agents with the capacity to be interpersonally reasonable rather than as patients narrowly interested in having their individual needs met. Sen makes the distinction between rationality (reasoning with oneself) and reasonableness (reasoning with others) (Sen, 2011, p. 197; see also Bartkowski & Lienhoop, 2018), arguing that the latter is a more compatible reasoning mode with a social choice approach to collective decision-making, in particular the idea that "a reasonable action is one that can be sustained in the light of the need to provide *impartial* reasoning" (Sen, 2011, p. 9).
3. Supporting institutions. Finally, ensuring the capacity for citizens to participate in such a deliberative process, as well as ensuring the legitimacy of and adherence to decision-making outcomes, requires the creation and maintenance of supporting institutions. This would involve, for example, changing incentive and other regulatory structures that impede the practice of social choice, such as the institutionalized requirement to prioritize the metrics of cost-benefit analysis in many public coastal management projects.

A large amount of work on capabilities and development has been inspired by Sen, even specifically within the context of sustainability (Pelenc & Ballet, 2015; Martins, 2013; Ballet et al., 2011). Discussion of the decision-making side of Sen's capabilities approach has however remained surprisingly scant. The most relevant discussion would probably be that of Bartkowski and Lienhoop (2018), who selectively draw on Sen's philosophical and welfare economic work to strengthen the theoretical foundation of deliberative forms of monetary valuation of environmental assets. They recognize deliberative monetary valuation as an "imperfect information tool" that "does not replace a political decision-making process ... [but informs it]," but seek to enrich it anyway "for lack of better alternatives" (p. 102). I would argue that social choice is itself a better alternative in its own right, but little work to date has focused on developing social choice as a viable alternative, and even less work has focused on how to actually achieve it in practice.

There are, of course, a variety of mechanisms available for instigating social change, but not all would support the adoption of social choice. Relying on market mechanisms through economic choice, for example, has reductionist informational requirements that would never—indeed, could never—provide the justification for the adoption of a kind of capability-based social choice evaluative procedure that has no a priori allegiance to economic logic. This is because cost-benefit analysis, by necessity, must translate all considerations into a unitary metric (money), which in turn homogenizes value and thus makes it impossible to consider values other than monetary value. Relying on government legislative action through traditional political processes such as voting, too, remains limited to the individually focused logic of majority rule (see Sen, 1999). In light of the limitations of economic logic or formal government regulation to address persistent environmental problems, the next place to look for the capacity to instigate social change is civil society, a source of political power capable of exerting pressure on government and market institutions (usually getting to the latter through the former) outside of the normalized courses of action (Smith et al., 2020). This means that, by engaging in collective action, civil society organizations promote, often successfully, their shared interests even if these interests are, for example, deemed undesirable from the perspective of economic growth, or are practically impossible to fulfill through established means of political reform.

Movements in civil society can target or spread to a variety of levels of social organization when making claims for social change, a process Tarrow (2011, pp. 192–193) calls "scale shift," meaning the spreading of a contentious issue and its shifting "to levels of the polity in which new opponents, new potential alliances, and different institutional settings" begin to shape the movement's progress. This has historically been the case for a variety of social movements. Tarrow (2011, p. 239) notes, for example, how the percentage of international civil society organizations focused on environmental issues has increased from 6% in the 1950s to nearly 20% in the 2000s. Indeed, the kinds of institutional change that would create conditions conducive to practicing social choice in environmental evaluation would need to be implemented at a variety of administrative levels, from federal to state to local. This would require building coalitions across communities of interest, for instance, linking concerns in the case of Flagler Beach with overlapping interests in Flagler County, Florida, and other US communities. It is often the case that, if pressure for change is targeted only at local levels, then higher-level prioritization will continue to trump priorities established at lower levels of organization. I now turn to discuss these challenges through the example of a sustained and ultimately unsuccessful civil society campaign in Flagler Beach, before expanding the discussion to the challenges that face local movements more broadly.

Local organizing for the environment: The rise and fall of Save Flagler's Beach

While many actors have been involved in the decades-long debate over coastal management in Flagler Beach, one local organization, named Save Flagler's Beach (SFB), stands out as perhaps the sole civil society group to be intimately involved in the environmental conflict over the years. SFB, a group the town mayor described as a "very, very, very vocal group of just local citizens" (FBM, Interview 1), was founded around 2004. Largely initiated and coordinated by two women, both residents of Flagler Beach, the group eventually maintained an active volunteer base of some two dozen concerned citizens who regularly advocated for their cause at public meetings, church gatherings and private events. Over the course of nearly 10 years, SFB sustained an extended campaign of vocal opposition to a variety of proposed erosion control strategies in Flagler Beach and Flagler County, while simultaneously promoting what they saw as the most desirable alternative.

The central claim made by SFB participants was that the strategies being proposed by FDOT and USACE would not be effective at reducing erosion, or more generally, would not address local citizen concerns over environment and recreation (Potter, 2011). The normal course of decision-making, they argued, was leading local government astray. A marginal technology trademarked as an "undercurrent stabilizer" by Holmberg Technologies, Inc., they insisted, would instead lead to the most effective restoration of the dunes and vegetation, thus providing the desired environmental, recreational and economic benefits (Cavaliere, 2012). SFB members became familiar with the Holmberg technology after the company had lobbied in the town for several years, including in online forums where many SFB members were first exposed to the "undercurrent stabilizer" concept. These "undercurrent stabilizers" were claimed to restore sand to a beach without the need for large dredging operations or construction equipment to grade the restored beach. Rather, Holmberg Technologies maintained that its technology redirects natural subsurface currents that transport sediment to deposit their load on eroded beaches rather than letting it be transported along the coast. The implication of adopting this technology was that both the renourishment project advocated by USACE and the seawall advocated by FDOT would no longer be necessary, supposedly saving the town money and political autonomy.

In a personal communication (FBR, Interview 9), one of the founding members of SFB explained how the group's lobbying efforts were coordinated primarily by two local residents, who then had around two dozen active volunteers engaged in outreach and other activities within the community. They also described the various techniques used by the group to promote their interests, what Tilly (2008) might call the "repertoire of contention," which involved maintaining a website, attending and speaking at public meetings, visiting and holding speeches at churches and women's clubs, petition drafting and submission, and information outreach at community events. All these activities focused on promoting the Holmberg technology while vilifying other proposed projects. The SFB group also provided informational packages promoting the Holmberg technology to decision-makers.

Early on, SFB sought to increase its visibility and credibility by lobbying local residents to become group supporters and participants, recruiting, for example, members from local churches and women's groups, as well as gaining the trust and support of several local elected officials. One Flagler Beach commissioner self-reported having "become very familiar with Holmberg's technology" and spending "copious hours with the Save Flagler's Beach group" (FBCC, Interview 2). SFB often had a prominent presence at public meetings and project workshops, where SFB volunteers and other supporters of the Holmberg technology publicly questioned the desirability of USACE and FDOT involvement while promoting the Holmberg technology (FlaglerLive, 2011). SFB members were also very active in the online comment sections of local news reports related to beach management in general and the Holmberg technology in particular.

SFB's persistence demonstrated a strong commitment to its cause, with group members fiercely contesting any criticism brought against the Holmberg technology and continuing to push for adoption of the technology even in the face of strong resistance by USACE, FDOT, Flagler County and local officials. One county official (FCPZD, Interview 4) referred to SFB as "folks with very strong opinions, who effectively became the vocal minority," recalling them as "the ones who get the time in front of the elected officials, and then have that almost exclusive audience." Similarly, a Flagler County engineer, when asked about SFB's presence at public meetings, told me that "there are some who just so strongly supported Holmberg and that sort of system, and it's all they could talk about, it's all they could think about" (FCE, Interview 8).

After nearly 10 years, the persistence of the SFB group led the Flagler County Tourism Development Council to grant $50,000 in funding for a preliminary study to assess the feasibility of employing a project by Holmberg Technologies (Hoye, 2012). The project was, nevertheless, eventually rejected unanimously by the Flagler Beach City Commission in March 2013 (Overstreet, 2013), as they considered the resulting study to be of insufficient quality in a variety of ways (see Hoye, 2013; FlaglerLive, 2013a, 2013b). One town official, for example, called Holmberg's product "snake oil," sarcastically telling me in an interview that "[Holmberg] had a cure for everything" (FBPWD, Interview 3). Not all officials were so critical, however. For example, one Flagler County official explained his reluctance to support the Holmberg technology this way: "I can't say that the Holmberg stuff alone was flawed, I think that our perspective was just that it wasn't tested enough. And that even in his defense of the science that the examples they provided were few and not local. And the idea that we had with it was it had merit if it ... had been vetted through other agencies." (FCPZD, Interview 4). Among other problems, the technology had been extensively and heavily criticized by academics and government agencies, both due to accusations of misleading claims made by company representatives and evidence that the technology has been ineffective in trial installations (for example, Pilkey & Sampson, 1998).

After nearly 10 years of lobbying efforts by SFB, the Holmberg alternative was officially taken off the table by the Flagler Beach City Commission. The failure of the SFB group's efforts may have had important repercussions beyond the group itself. In particular, some have suggested that it may have reduced the possibility for similar kinds of civil society contestation in the future. For example, a Flagler County official told me that after the rejection of the Holmberg proposal, its advocates "felt that 'well we made the attempt and now this is what we have for this 10-year effort ... really nothing to show for it'" (FCPZD, Interview 4). This is reminiscent of the potential, noted by Tilly (1977, p. 24), of failed campaigns in civil society to "both use up the available resources and reduce the willingness of individual members to commit themselves to the collective effort" in the future.

SFB and the wider challenge of organizing for the environment

In the local context of Flagler Beach, SFB demonstrated the potential for mobilization of residents and resources over an extended period of time in pursuit of an alternative that under the circumstances could only be introduced through collective civil society pressure, rather than being the result of formal economic or political choice procedures. They also demonstrated the importance of being organized and ready to take advantage of "political opportunities," which SFB did in order to secure the $50,000 tourism development grant, which only became possible after the commission rejected an alternative proposal (the FDOT seawall) and a more uncertain alternative was left on the back burner (USACE renourishment). However, the group's effectiveness was clearly quite limited, amounting in the final analysis to what one Flagler County official called "noise" and only a "temporary divergence" from the normal course of decision-making (FCE, Interview 8). The three major problems that seemed to affect the efficacy and legitimacy of SFB mirror more general challenges to progressive collective action throughout the state and country.

First of all, while SFB was loud and proud, they were and remained a "vocal minority." In fact, the vast majority of Flagler Beach citizens did not engage in the debate over the erosion issue and did not support SFB. In other words, SFB was unable to expand its base, either by expanding its membership or through alliance building. The second problem is related to the fact that SFB tended to target local authorities (city and county) who lack the kind of decision-making power necessary to permit such a project, while higher-level authorities remained unaffected in their categorical opposition. In other words, SFB aimed their efforts at the wrong target or at least emphasized local targets at the expense of relevant targets at higher levels of organization. Finally, the group was ultimately promoting a scientifically dubious technology rather than a strategy based on or supported by scientific best practice. Trial installations of the technology in other parts of Florida and the US had regularly resulted in their ordered removal by state agencies due to malfunctioning, rapid degradation or even exacerbation of erosion problems. Many state and federal agencies had denied the technology installation permits, and some renowned coastal scientists argued directly that the Holmberg technology was based on deceptive and misleading descriptions of coastal sedimentary processes (Pilkey & Sampson, 1998). The resistance of SFB members to this criticism was, however, to a large degree rooted in a distrust of "experts" and politicians who, according to one SFB member (FCR, Interview 9), lacked the "guts" to stand behind a project which "interfered" with the status quo, suggesting instead that "money, power and greed" had won the day. In other words, SFB mobilized problematic knowledge resources in their campaign, which speaks to the importance of the appropriateness of the knowledge movements draw on to advance their causes. The same kinds of problems that limited the SFB efforts in Flagler

Beach can be identified at a more general level in American civil society, in particular challenges around expanding movement support and building alliances, identifying appropriate targets for movement campaigns and the need for adequate knowledge and other material resources.

Alliances: The need for an expanded base

SFB, while garnering some support at the local level, was never able to expand its base beyond Flagler Beach or the specific issue of erosion control technology. One of the main arguments made by SFB was that the Holmberg technology would be not only a more cost-effective approach to erosion management, but a much more environmentally friendly alternative as well, which remained a central pillar of SFB's campaign for years. However, this environmental framing may not appeal to as wide an audience as anticipated, particularly when we consider that a significant portion of the American public are either apathetic about the environment or hold explicitly anti-environmental sentiments.

According to a public opinion survey conducted by the Associated Press-NORC Center for Public Affairs Research and the Yale School of Forestry and Environmental Studies (2015), nearly 60% of the American adult population is either apathetic about environmental issues or actively "anti-environmentalist" to some degree, meaning they, for example, deny there is an environmental crisis, express no interest in the topic or insist that environmental degradation is not problematic. In Flagler Beach, the majority (62%) of citizens who completed a beach management public opinion survey conducted as part of my dissertation research in 2016 (n = 369, Boda, 2018a, Appendix 1) rated the quality of the local beach environment as "excellent," while only 3% rated it as "below average," suggesting the idea that the town has an environmental crisis is not particularly widespread among the public (as a comparison, the federal government considers more than 80% of Flagler County's coastline to be critically eroded). However, the problem is more than one of awareness or education. In Flagler Beach, for example, some citizens, rather than lacking knowledge about the issue of environmental conservation, have instead expressed their frustration with what they see as excessively restrictive environmental regulations in the town, writing open comments in the 2016 public opinion survey such as: "I am sick and tired of all the turtle Nazis. Flagler Beach is for people also." And: "Too many rules regarding turtles, they're fine, and besides our beach in comparison to the rest of Florida is relatively nil as in turtle nesting."

Rather than prioritizing environmental conservation, many citizens argue that higher priority should be placed on economic growth or jobs. Some Flagler Beach residents, for example, complained in open comments that "Flagler Beach does not look out for local business" or that they "do not begin to understand the lack of business minded city leaders." In the US more generally, adults ranked the economy as the second-highest priority for the nation in a 2017 Pew Research Center survey, and economy has remained a top priority for at least the last decade, particularly since the 2007 financial crisis (Aisch & Parlapiano, 2017). Another 2017 Pew Research Center survey (Anderson, 2017) found that nearly one in four adults think that "the country has gone too far in its efforts to protect the environment," with the divide between advocates and opponents increasingly along political partisan lines. Nearly three in five Republicans and Republican leaners (58%), for example, said in the 2016 Pew Research Center survey (Bialik, 2016) that environmental laws and regulations cost too many jobs and hurt the economy, a huge increase from 34% in 2007. This trend is made more relevant when we are reminded of the fact that Florida's state government has been dominated by Republicans for decades. Regardless of its distribution along party lines, the lack of prioritization of the environment is widespread, with roughly 45% of American adults in 2017 saying that the environment should not be a top priority (Anderson, 2017). In the face of such resistance by both citizens and political parties, local environmental movements will likely need to expand their base of support beyond environmentalists and environmental messaging to include alliances with movements advocating for other pressing social and economic grievances, if they are to gain traction and effect change (see O'Byrne, 2020).

Targets: The need to go beyond the local level

Even if people do believe the environment needs timely protection, commonly held or promoted perspectives regarding the appropriate scale to address environmental problems may pose a problem. What has been termed the "local trap," that is, the widely held assumption that "localized decision-making is inherently more socially just or ecologically sustainable" (Purcell & Brown, 2005, p. 279) has become a sort of common sense in public environmental discourse. As we have seen, most movement activities carried out by SFB were directly targeted at local government authorities, in particular, the Flagler Beach City Commission, which is the primary municipal decision-making body. The common sense thinking here is that erosion along State Road A1A is a local problem, so the local government should fix it. However, the argument in support of a localist approach to addressing environmental problems runs into two primary challenges. The first has to do with the practical power relations between hierarchical levels of government administration, and thus which agencies actually have formal legal capacity to address a particular problem. The second has to do with the well-established principles of scientific best practice derived from conservation research, which challenge the notion that keeping a local focus is desirable or sufficient for addressing many pressing environmental problems.

The first argument for targeting change efforts at local governments is that they are the most immediate, democratically accountable public body with the power to make decisions about public property, including environmental management, and this is in many ways true. Flagler Beach does in fact have substantial autonomy over many of its local environmental resources. However, there are important limits to the extent of local government's competence. In fact, the city commission sits at the lowest level of an administrative hierarchy that spans from local to county to state to federal government, and as such it must abide by the rules and regulations imposed by all these higher-level government bodies (see Boda, 2018a, Section 3.2). These higher-level constraints are particularly relevant when it comes to erosion management along State Road A1A. This is because the FDOT has primary responsibility for maintaining this state road, not Flagler Beach, and the FDOT's right-of-way encompasses the most severe erosion areas within city limits, meaning that, in the areas where erosion is most severe, the city lacks any legal authority (Boda, 2015). Furthermore, the projects SFB opposed, namely the FDOT seawall and the USACE renourishment project, were all products of higher-level government agencies over which Flagler Beach government has very little meaningful influence. Due to these preexisting administrative hierarchies, SFB, or any local movement for that matter, would need to target their efforts not only at local government, but also those agencies with the legal capacity to address the issue or change the content of proposed projects.

The second argument for targeting local government is that local residents necessarily have the know-how and incentives to sustainably manage environmental resources. A Flagler Beach resident, for example, expressed this popular opinion in an open comment in the 2016 public opinion survey: "If it wasn't for the bureaucracy of government regulations, we could repair the dunes ourselves. Locals would know not to use little rocks and paper or seawalls to repair the dunes. Income from meters, tourists, taxes should be used for maintenance. Keep responsibility local and never cede management to county or state."

However, despite the insistence by some that the more local, the better, it is canon in conservation science that the effective, long-term management of ecological systems requires the protection of large areas of maximum diversity that are well-connected (i.e., have minimal fragmentation) and have intact interiors (i.e., minimal edge) (Haddad et al., 2015). These requirements in turn necessarily involve cross-boundary coordination within and between levels of social organization, and indeed many prominent conservationists have called for just this kind of multilevel collaboration and coordination ranging from the local to the global (Rands et al., 2010). Adding to this challenge of cross-scale coordination, the tendency toward localization in movements, which Tarrow (2011, pp. 131–132) discusses under the rubric of the "tyranny of decentralization," can have detrimental implications for the longevity and coordination of collective action within civil society: "While encouraging the autonomy of the base and exhilarating activists with a sense of participation, [decentralized movements] permit—and indeed encourage—a lack of coordination and continuity."

Mobilizing structures: The need for adequate material and knowledge resources

In addition to building alliances and targeting the appropriate authorities, movements need to be able to draw on adequate knowledge and material resources. Adequate knowledge resources, for example, provide a way to legitimize the movement's agenda, support their efforts with valid evidence and inform their actions, while material resources such as funding, labor and technology ensure the movement has the means to continuously engage in advocacy activities over the long term. Given the size of SFB, its use of material resources was minimal, and largely based on volunteer labor and contributions in kind from active members (e.g., in designing websites, printing materials for meetings, etc.). If, as discussed above, SFB had built alliances with other movements and attempted to target higher-level authorities, its need for material resources would surely have grown. When it comes to knowledge resources, however, we have seen how SFB put all its eggs into one technological basket, which upon closer inspection proved a dubious technology without support from either the scientific community or relevant government agencies, ultimately undermining its efforts and delegitimizing the movement in the eyes of some local and county authorities. Clearly not every citizen can be an expert in coastal erosion technology; however, the resistance of SFB members to arguments that the Holmberg technology was potentially problematic help point toward a larger challenge facing movements for the environment. Perhaps one major, overarching challenge in this regard is the loss of legitimacy of scientific knowledge and expertise more generally, both among the public and in the academy itself. The important point is not so much that scientists know exactly what to do, but that objective knowledge of political and economic systems, for example, is indispensable for the effective actions of collective agents (Boda et al., 2022a; Boda & Jerneck, 2019; Isgren et al., 2019).

A Pew Research Center survey (Funk & Kennedy, 2016), for example, found that there are more American adults who believe that scientific findings are influenced by scientists' desire to advance their personal careers (36%) than there are who believe that scientific findings are based on the best available scientific evidence (32%), although there are strong differences in opinion along partisan lines. The same survey showed that, regardless of political affiliation, more than 20% of American adults have little to no confidence that scientists act in the best interest of the public. Such a "legitimacy crisis" of science has deeper causes in what Faran and O'Byrne label "epistemological cynicism," meaning a skeptical view of, or outright denial of, the possibility of producing objective scientific knowledge, based in what critics see as "the failure of reason to deliver the promise of modernity and indeed, in some formulations, its complicity in the disasters of modern times, from colonialism, to genocide and environmental destruction" (Faran & O'Byrne, 2015, p. 3). "It would be simplistic," the authors clarify, "to put down the legitimacy crisis of science to epistemological cynicism. The practice of science as a social institution must shoulder its share of the blame," pointing in particular to the tendency of some academics to maintain a naïve understanding of the role of scientific knowledge in social change, the tendency for others to collapse all knowledge types into the realm of relativist subjectivity and, more broadly, to troubling tendencies toward the commercialization of academic institutions and the resulting impacts on research funding allocation and quality control criteria.

Arguably a crucial, yet overlooked aspect of how movements draw on knowledge resources has to do with the forms of interaction between civil society organizations and the practice of scientific knowledge production. While many movements may rightfully resist external "experts" telling them how to conduct their affairs, the lack of any interaction between scientific knowledge producers and civil society organizations runs the risk of movements drawing on faulty analyses to inform their campaigns. An alternative to the end-of-pipe approach to movements drawing on science could be based around a continuous, reflexive and critical dialogue between movement participants and scientific knowledge producers, such that the knowledge being picked up by movements can be collaboratively assessed and, when necessary, improved upon. Nurturing a more fruitful relationship between civil society and scientific knowledge production will require major changes to the normal conduct of research as carried out in institutions of higher education, not least in the realm of research funding and criteria for what is considered desirable "impact."

Conclusion

The theoretical approach adopted in pursuit of sustainable development has immense repercussions for the practical activities that will be deemed necessary and desirable to achieve it, including the kind of decision-making procedures that should be applied when faced with difficult trade-offs in practice. Different approaches lead to different practical strategies, which leads to tension and disagreement among advocates of these competing approaches. The only way to settle such disputes is through political advocacy, engagement and struggle. I have argued that social choice, a decision-making procedure developed within the capabilities approach to human development, with its focus on identifying contextually relevant decision criteria through reasoned public deliberation, holds great promise for overcoming the limitations of decision-making focused solely on economic efficiency and cost-benefit analysis. However, despite its promises, social choice will not become a reality on its own, but must be struggled for deliberately—in particular, through social movements that emanate from civil society. However, as I attempted to show using the case of SFB, these movements, particularly at the local level, face many challenges, including how to expand their base and build alliances, where to target their efforts and what kind of knowledge resources to draw on when strategizing the movement's activities. One fruitful way forward may be to explore more radical forms of collaboration between civil society organizations and scientific knowledge producers, such that the tactical and other forms of experiential knowledge, so essential to movements, can be combined with objective knowledge of economic and political structures. The correct application of actionable scientific knowledge can, for example, aid in the identification of objective interest alliances and in the targeting of appropriate authorities (O'Byrne, 2020; Isgren et al., 2019; Harnesk and Isgren, 2021). Overcoming skepticism of the relevance of science; however, will require opening both space and resources for new forms of research collaborations, as well as practical demonstration of the usefulness of scientific analysis for civil society movements.

References

Aisch, G., & Parlapiano, A. (2017, February 27). "What do you think is the most important problem facing this country today?" *New York Times*. https://www.nytimes.com/interactive/2017/02/27/us/politics/most-important-problem-gallup-polling-question.html

Allen, A. (2012). The unforced force of the better argument: Reason and power in Habermas' political theory. *Constellations, 19*(3), 353–368. https://doi.org/10.1111/cons.12005

Anderson, M. (2017, April 20). For Earth Day 2017, here's how Americans view environmental issues. Pew Research Center. https://www.pewresearch.org/fact-tank/2017/04/20/for-earth-day-heres-how-americans-view-environmental-issues/

Associated Press-NORC Center for Public Affairs Research & Yale School of Forestry & Environmental Studies. (2015). *Public opinion and the environment: The nine types of Americans*. Issue brief. https://apnorc.org/wp-content/uploads/2020/02/12-2015-Segmentation-Report_D10_DTP-Formatted_v2b-1b.pdf

Ballet, J., Bazin, D., Dubois, J.-L., & Mahieu, F.-R. (2011). A note on sustainability economics and the capability approach. *Ecological Economics, 70*(11), 1831–1834. https://doi.org/10.1016/j.ecolecon.2011.05.009

Bartkowski, B., & Lienhoop, N. (2018). Beyond rationality, towards reasonableness: Enriching the theoretical foundation of deliberative monetary valuation. *Ecological Economics, 143*, 97–104. https://doi.org/10.1016/j.ecolecon.2017.07.015

Bialik, K. (2016, December 14). Most Americans favor stricter environmental laws and regulations. Pew Research Center. https://www.pewresearch.org/fact-tank/2016/12/14/most-americans-favor-stricter-environmental-laws-and-regulations/

Boda, C. S. (2015). Power and rationality in coastal planning: Effects on participation and possibility in the management of barrier island dunes in Flagler Beach, Florida, USA. *Journal of Coastal Conservation*, 19(4), 561–576.

Boda, C. S. (2018a). *The beach beneath the road: Sustainable coastal development beyond governance and economics*. PhD thesis, Lund University.

Boda, C. S. (2018b). The entrepreneurial Sunshine State: Neoliberalism, growth management and environmental conservation in Florida. *Journal of Urban Affairs, 40*(6), 838–862. https://doi.org/10.1080/07352166.2017.1413287

Boda, C. S. (2018c). From economic choice to social choice in coastal management: A critical assessment of the use of cost-benefit analysis in the evaluation of an erosion control project in Flagler County, Florida, USA. *Ocean & Coastal Management, 162*, 85–99. https://doi.org/10.1016/j.ocecoaman.2017

Boda, C. S., & Faran, T. (2018). Paradigm found? Immanent critique to tackle interdisciplinarity and normativity in science for sustainable development. *Sustainability, 10*(10), 3805. https://doi.org/10.3390/su10103805

Boda, C. S., & Jerneck, A. (2019). Enabling local adaptation to climate change: Towards collective action in Flagler Beach, Florida, USA. *Climatic Change, 157*, 631–649. https://doi.org/10.1007/s10584-019-02611-6

Boda, C. S., Faran, T., Scown, M., Dorkenoo, K., Chaffin, B. C., Nastar, M., & Boyd, E. (2021). Loss and Damage from Climate Change and Implicit Assumptions of Sustainable Development. *Climatic Change, 164*(13). https://doi.org/10.1007/s10584-021-02970-z

Boda, C. S., O'Byrne, D., Harnesk, D., Faran, T., & Isgren, E. (2022a). A collective alternative to the inward turn in environmental sustainability research. *Journal of Environmental Studies and Sciences, 12*(3), 291–297. https://doi.org/10.1007/s13412-021-00738-6

Boda, C. S., Scown, M., Faran, T., Nastar, M., Dorkenoo, K., Chaffin, B. C., & Boyd, E. (2020). Framing loss and damage from climate change as the failure of sustainable development. *Climate and Development, 13*(8), 677–684. https://doi.org/10.1080/17565529.2020.1851640

Boda, C. S., Scown, M. W., & Faran, T. (2022b). Forgotten coast, forgotten people: Sustainable development and disproportionate impacts from Hurricane Michael in Gulf County, Florida. *Natural Hazards, 111*, 877–899. https://doi.org/10.1007/s11069-021-05082-0

Cavaliere, M. (2012, May 2). City OKs stabilizers in Flagler Beach. *Palm Coast Observer.* https://www.palmcoastobserver.com/article/city-oks-stabilizers-flagler-beach

Faran, T., & O'Byrne, D. (2015). Like poets in times of dearth: The legitimacy crisis of science and social movements. Paper presented at the First International Conference in Contemporary Social Science, Rethymno, Greece

FlaglerLive. (2013a, February 12). Cut-and-paste: Holmberg Beach study falls short of city commissioners' expectations. *FlaglerLive.com.* https://flaglerlive.com/50800/dick-holmberg-stabilizers/

FlaglerLive. (2013b, February 15). An insulted Flagler Beach Commission rebuffs Holmberg's plan and wants its money back. *FlaglerLive.com.* https://flaglerlive.

com/50900/an-insulted-flagler-beach-commission-rebuffs-holmbergs-plan-and-wants-its-money-back/

FlaglerLive. (2011, September 20). In Flagler Beach, questions, ridicule and anger in search for beach-saving answers. *FlaglerLive.com*. https://flaglerlive.com/28401/flagler-beach-town-hal/

Funk, C., & Kennedy, B. H. (2016, October 4). The politics of climate. Pew Research Center. https://www.pewresearch.org/science/2016/10/04/the-politics-of-climate/

Gardoni, P., & Murphy, C. (2008). Recovery from natural and man-made disasters as capabilities restoration and enhancement. *International Journal of Sustainable Development and Planning, 3*(4), 317–333. https://doi.org/10.2495/SDP-V3-N4-317-333

Gardoni, P., & Murphy, C. (2009). Capabilities-based approach to measuring the societal impacts of natural and man-made hazards in risk analysis. *Natural Hazards Review, 10*(2), 29–37. https://doi.org/h10.1061/(ASCE)1527-6988

Gardoni, P., & Murphy, C. (2010). Gauging the societal impacts of natural disasters using a capability approach. *Disasters, 34*(3), 619–636. https://doi.org/10.1111/j.1467-7717.2010.01160.x

Haddad, N. M., et al. (2015). Habitat fragmentation and its lasting impact on Earth's ecosystems. *Science Advances, 1*(2), e1500052. https://doi.org/10.1126/sciadv.1500052

Harnesk, D., & Isgren, E. (2021). Sustainability as a real utopia: Heuristics for transformative sustainability research. *Environment and Planning E: Nature and Space, 5*(3), 1678–1695. https://doi.org/10.1177/25148486211018570

Hoye, M. (2012, November 7). Holmberg contract approved, *Palm Coast Observer*. https://www.palmcoastobserver.com/article/holmberg-contract-approved-o

Hoye, M. (2013, February 15). Provencher to Holmberg: "Shame on you." *Palm Coast Observer*. https://www.palmcoastobserver.com/article/provencher-holmberg-shame-you

International Union for the Conservation of Nature. (2017). *Everglades National Park: Conservation outlook*. IUCN World Heritage Outlook.

Isgren, E., Boda, C. S., Harnesk, D., & O'Byrne, D. (2019). Science has much to offer social movements in the face of planetary emergencies. *Nature Ecology and Evolution, 3*, 1498. https://doi.org/10.1038/s41559-019-1024-x

Martins, N. O. (2013). The place of the capability approach within sustainability economics. *Ecological Economics, 95*, 226–230. https://doi.org/10.1016/j.ecolecon.2013.07.004

O'Byrne, D. (2020). A contribution to building unified movements for the environment: aligning interests, forming alliances. *Human Geography, 13*(2), 127–138. https://doi.org/10.1177/1942778620927391

Overstreet, P. (2013). Regular meeting of the Flagler Beach City Commission, Thursday, March 14, 2013, at 5:30 P.M. Flagler Beach City Commission.

Pelenc, J., & Ballet, J. (2015). Strong sustainability, critical natural capital and the capability approach. *Ecological Economics, 112*, 36–44. https://doi.org/10.1016/j.ecolecon.2015.02.006

Pew Research Center. (2017, January 24). After seismic political shift, modest changes in public's policy agenda. https://www.pewresearch.org/politics/2017/01/24/after-seismic-political-shift-modest-changes-in-publics-policy-agenda/

Pilkey, O. H., & Sampson, D. W. (1998). White Paper: Notes on the Holmberg Undercurrent Stabilizer™ program for the study of developed shorelines, Duke University.

Potter, T. (2011, September 13). Seawall and "renourishment" alternative: Saving the beach without losing a town's soul, *FlaglerLive*. https://flaglerlive.com/28090/gc-flagler-beach-seawall-dunes/

Purcell, M., & Brown, J. C. (2005). Against the local trap: Scale and the study of environment and development. *Progress in Development Studies, 5*(4), 279–297. https://doi.org/10.1191/1464993405ps1220a

Rands, M. R., et al. (2010). Biodiversity conservation: Challenges beyond 2010. *Science, 329*(5997), 1298–1303. https://doi.org/10.1126/science.1189138

Sen, A. (1999). The possibility of social choice. *American Economic Review, 89*(3), 349–378. https://doi.org/10.1257/aer.89.3.349

Sen, A. (2001). *Development as freedom.* Oxford University Press.

Sen, A. (2011). *The idea of justice.* Harvard University Press.

Smith, S. R., Christie, I., & Willis, R. (2020). Social tipping intervention strategies for rapid decarbonization need to consider how change happens. *Proceedings of the National Academy of Sciences, 117*(20), 10629–10630. https://doi.org/10.1073/pnas.2002331117

Tarrow, S. G. (2011). *Power in movement: Social movements and contentious politics.* Cambridge University Press.

Tilly, C. (1977). Studying social movements/studying collective action. CRSO Working Paper no. 168. Center for Research on Social Organization, University of Michigan.

Tilly, C. (2000). *Contentious performances.* Cambridge University Press.

Interviews (for full list of dissertation interviews, see Boda, 2018a, p. v)

1. Flagler Beach Mayor, FBM, January 16, 2014.
2. Flagler Beach City Commissioner, FBCC, January 16, 2014.
3. Flagler Beach Public Works Director, FBPWD, March 11, 2015.

4. Flagler County Planning and Zoning Director, FCPZD, March 12, 2015.
8. Flagler County Engineer, FCE, March 18, 2015.
9. Flagler Beach resident and SFB founding member, FBR, September 17, 2015.

About the author

Chad Boda is senior lecturer in environmental science at the Department of Urban Studies, Malmö University. He is an interdisciplinary geographer with a focus on environmental conservation and sustainable development. His current research is a collaboration between Sweden, Zimbabwe, Uganda and Ghana studying the emergence, development and outcomes of farmer-based rural social movements for sustainable agriculture in sub-Saharan Africa.

11. Practices of resilience

Questioning urban adaptation in the Chilean social upsurge

Cristina Visconti

Abstract
Technocratic and market-driven logics that inform planning systems and urban adaptation often exacerbate socio-spatial and climate inequalities. Facing this challenge, grassroots social movements build their own bottom-up resilience practices to contest neoliberal urban spaces. Building on her ethnographic work in the *ecobarrios* (eco-neighborhoods) of Chile, Visconti explores a case in which community actions were catapulted by the sociopolitical turmoil and the counter-events held during COP25 in Santiago in December 2019. She observes that while *ecobarrios* have a huge potential for transformation, such transformation can only materialize insofar as they mobilize the right resources and avoid the "local trap."

Keywords: *buen vivir*, *ecobarrios*, socio-territorial vulnerability, grassroots resilience, Chile

In November 2019, Chile, the host country for the COP25 summit, suspended the event after weeks of street protest against the socio-environmental impacts of the extreme neoliberal policies that affect the most vulnerable and disadvantaged groups of population. In cities, this results in socio-spatial and climate inequalities exacerbated by technocratic and market-driven logics that inform planning systems and urban adaptation. The social upsurge and the call for a new constitution to guarantee social justice and the protection of natural resources as a commons have intersected grass-roots climate initiatives with decades-old territorial socio-environmental demands challenging institutional climate discourse. This chapter discusses how bottom-up resilience practices, such as the Chilean *ecobarrios*, create alternatives to neoliberal climate agendas in contested urban spaces. The

Armiero, M., E. Turhan & S. Paolo De Rosa (eds), *Urban Movements and Climate Change: Loss, Damage and Radical Adaptation*. Taylor & Francis Group 2023
DOI: 10.5117/9789463726665_CH11

case also demonstrates how the emerging Latin American debate about the rights of nature and *buen vivir* (living well) can influence urban adaptation through a postcolonial perspective.

Social upsurge and the COP25 counter-events

Chile has become an international example of economic development in the past five years, lowering its poverty levels from 38.6% in 1990 to 8.6% in 2017 (PNUD, 2018). As the first Latin American member of the OECD, Chile has one of the fastest-growing economies of the last decades (World Bank, 2019). This economic growth narrative has been instrumental in building the nation's image as the "Oasis of Latin America," where the apparent financial wealth and relative political stability since the end of Augusto Pinochet's dictatorship in 1990 have became commonplaces in national political discourses. A few days before the intense social turmoil of October 2019, Sebastián Piñera, at that time president of the Chilean Republic, used the *oasis* metaphor to justify the view of the country as the leading regional actor in the global response to climate change, a view that was supported by the Chilean presidency and hosting of COP25 (Navarro & Tromben, 2019; Poo, 2021).

Chile is not a major CO_2 emitter; it accounts for a mere 0.3% of global CO_2 emissions (Global Carbon Project, 2018). However, the country is highly exposed to climate-related hazards (Sandoval et al., 2021; Kreft et al., 2017). Climate change scenarios are projecting an increase of already evident environmental critical processes such as temperature rise, reduced precipitation, desertification, drought, water scarcity, loss of biodiversity and increased climate-related extreme weather (Cepal, 2012; Santibáñez, 2016).

Unequal wealth distribution exacerbates Chile's vulnerability to these environmental threats (OECD, 2021). The poorest 50% of households owned just 2.1% of the country's net wealth in 2017, while the richest 1% accounted for 26.5%, more than a quarter of the countries' net wealth (ECLAC, 2018). Along with income inequalities, the manifestation of urban inequalities in socio-spatial segregation is a significant issue in Chilean cities (Azocar et al., 2007; Bordsdof et al., 2007; Fernández et al., 2016).

The extractivist model dominated Chile during the 20th century, even before the radical neoliberalism installed by Pinochet's dictatorial regime fostered the role of the private sector as a provider of fundamental goods such as education, health, housing, water, energy provision and infrastructure (Correa-Parra et al., 2020; Sanzana Calvet & Castán Broto, 2016;

Sanzana Calvet & Castán Broto, 2015; Rodriguez et al., 2014; Holst et al., 2004; Elacqua, 2012; Posner 2012). Despite the transition to democracy, the neoliberal policies of Pinochet's political regime persist. As a result, economic overexploitation, socio-environmental conflicts, the spread of systemic and structural inequalities, segregation and violence have surged in the last 20 years (Fernández et al., 2016; Rodriguez et al., 2014). The consequences are increasing inequality, depletion of the commons and liberalization of the construction sector (Castán Broto & Sanzana Calvet, 2020).

On November 1, 2019, Chile, the host country for COP25, canceled the summit after weeks of street protests began on October 18, 2019 (18-O, see Figure 11.1). Chile's protests initially began over the price hike for subway tickets in Santiago, but they expanded, revealing the anger of ordinary Chileans, who felt they had been excluded from the nation's economic rise. The protests paralyzed the country and were inflamed by the declaration of a state of emergency, police repression and curfews. This social upsurge gave place to a movement of self-organized assemblies (*cabildos*) spread all over the country as a form of bottom-up action and advocacy responding to the severe sociopolitical crisis. The grassroots social and environmental agendas overlap became central to the protestors' main demand for a new constitution. This convergence was instrumental in bringing about the policies and actions of several grassroots movements. Among others, Cumbre de los Pueblos and Sociedad Civil por la Acción Climática (SCAC),[1] formed in response to COP25, attempted to outline alternative transformation pathways for climate adaptation based on communities and justice.

This chapter presents an investigation into the situation described above, which triggered grassroots debate and community actions animated by both the social turmoil and the participatory counter-events to the COP25 held in Santiago in December 2019. The social unrest influenced the grassroots climate agenda, and it made the topics of environmental and climate justice part of the discussion of a new social contract in Chile. The protesters affirmed in the street that the social crisis was also an environmental and climate-related crisis (Bouyè, 2019).

The chapter's ethnographic fieldwork consisted of participant observation and in-depth interviews that were conducted from September 2019 to

1 Cumbre de los Pueblos is an assembly of Latin American political organizations and social movements proposing a grassroots approach to climate change in opposition to the top-down decisions of COP25. Sociedad Civil por la Acción Climática (Civil Society for Climate Action, SCAC) is a platform that brings together more than 80 organizations, including environmental associations, territorial movements, professional associations, labor unions, and political and academic organizations. During COP25, SCAC elaborated a grassroots Declaration for Climate Emergency.

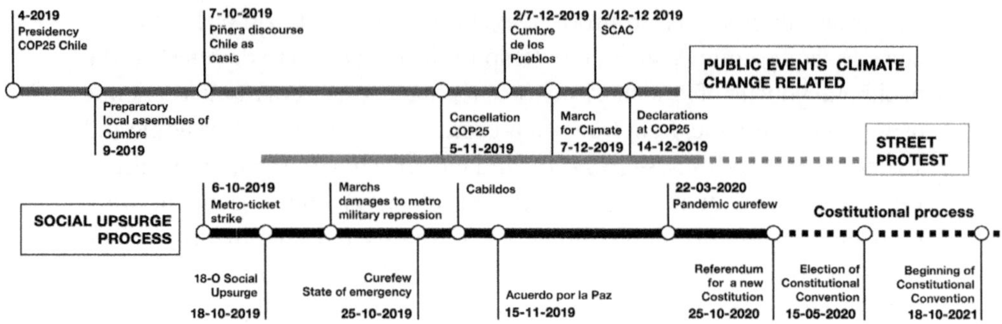

Figure 11.1. A timeline for the different junctures of the Chilean uprising. Source: Cristina Visconti.

January 2020. The author participated as an activist in the regional sub-assembly of Cumbre de los Pueblos–V Region Valparaiso, where activists, citizens and scientists discussed at the regional level the local effects of the climate crisis and their intersection with decades of socio-environmental conflicts. Participating in the ongoing street protests and the emerging popular assemblies happening in every neighborhood of Santiago provided firsthand experience of the movement, offering the opportunity to witness the making of a grassroots platform for socio-environmental justice. This active participation showed how citizens have a deep awareness of the harmful effects of the neoliberal model and depletion of the commons.

Additional fieldwork was conducted to study the bottom-up urban practices of the *ecobarrios* (eco-neighborhoods) at the forefront of creating alternative sustainable models for living and part of the Chilean Transition Network. This chapter explores the rise of socio-environmental demands of the new constitution, the relationships between the neoliberal planning system, the current business-as-usual model in Chilean urban climate adaptation and the grassroots experience of the *ecobarrios*. Its conclusions outline how these Chilean resilience practices emerge in vulnerable contexts and shape the counter-hegemonic Latin American debate about post-extractivist pathways with essential lessons for just urban adaptation through transformative and situated perspectives. The Chilean case offers insights into how a bottom-up resilience urban agenda can promote counter-neoliberal practices grounded on nonutilitarian values. These practices challenge the market-driven logic in urban greening that characterizes Chile's neoliberal urbanism and are reproduced by climate urbanism. The engaged citizens encountered in the ethnographic work cope, manage and navigate the impacts of neoliberal city and climate change in their everyday lives (Castán Broto & Sanzana Calvet, 2020). Through solidarity, mutual support, local food production and community education about sustainability and environmental justice they mitigate urban inequalities by reshaping the fabric of the city toward a resilience based on a process of "commoning."

Socio-environmental demands and a grassroots climate agenda

A year later, the protests of 18-O led to the demand for a new constitution, which culminated in a referendum on October 25, 2020. The "approve" side won, leading to a constitutional convention. In May 2021, the convention members were directly elected from both political parties' representatives and independent candidates to draft the new constitution. It is expected to be subject to a third vote for approval or rejection in September 2022. The composition of the convention—with a strong presence of environmental, indigenous and feminist activists[2]—represented groups that are usually excluded from the political system (Oyarzún Serrano, 2021; Politzer, 2021). The convention constituted a historical shift toward equity and inclusivity, with a focus on socio-ecological claims and the commons (Poo, 2021; Solís et. al., 2021; Marconi 2021).

The civil mobilization for the new constitution is seen as the key to including fundamental changes about the role of the state in defending environmental and human rights. At the core of the constitutional debate are principles of plurinationality, environmental justice, territorial sovereignty and protection of the commons pursuing a post-extractive transition (Gudynas, 2020). Many socio-environmental claims have gotten the public's attention, and the declarations provided by both the Cumbre de los Pueblos and the SCAC link the call for climate justice to socio-environmental conflicts and the systematic violation of human rights in Chile and most Latin American countries (Valenzuela-Fuentes et al., 2021; Svampa, 2012; Machado Aráoz, 2012). Chilean local communities have resisted the multiplication of extractive activities, the privatization of the commons and the creation of sacrifice zones (Valenzuela-Fuentes et al., 2021; Smart Larrain, 2019; Svampa 2012). Chile experienced 125 active socio-environmental conflicts (INDH 2021) and 1,113 socio-environmental protest events related

2 One-third of the elected candidates are declared environmental activists, with representatives from important civil organizations for environmental rights, such as Movimiento Agua y Territorios (MAT), Modatima-Protección del Medioambiente and two representatives from SCAC.

to mining, energy, real estate, accessibility, and air and water pollution from 2012 to 2017 (Allain, 2019; Castán Broto & Sanzana Calvet, 2020). Of these conflicts, 35% involved indigenous territories (INDH, 2021), and most of them concerned flexible legislation and controls over environmental issues. This is one of the pillars of the Chilean neoliberal model (Machado Aráoz, 2009), explicitly supported by the constitution through the privatization of natural resources such as minerals, water and land (Gudynas, 2021).[3] This socio-environmental mobilization focuses on the protection of the commons to overcome the utilitarian view of nature as a stockpile of commodities and strategic resources (Svampa, 2012; Gudynas, 2019). Thus, socio-environmental activism claims the commons as an imperative that requires keeping natural, social and cultural heritage-driven goods out of the market because their value transcends their market price (Svampa, 2012). This implicitly refers to the social and legal systems with values and traditions that give identities to communities and allow self-governance through shared organization, management and collective action (Bollier, 2008; Esteva, 2014; Barkin et al., 2016).

During the Cumbre de los Pueblos assembly several political ecology scholars[4] argued that climate justice in Latin America is interlaced with decades-old environmental justice challenges and with the debate about the recognition of nature as having rights. The discussion about the rights of nature is indeed based on a biocentric perspective, acknowledging nature beyond its economical, ecological and utilitarian value as a nonhuman subject. This represents a conceptual and legal overcoming of the exploitative logics typical of extractivism (Acosta et al., 2018; Gudynas, 2019 and 2021) and accumulation through dispossession of the commons (Harvey, 2003). Thus, the rights of nature unfold parallel to the rights of humans and not as an extension of human rights (Gudynas, 2019). They underlie an interdependent relationship between human and nonhumans based on respect and reciprocity (Acosta, 2019). This relationship encompasses an ecological justice perspective beyond the purely environmental perspective (Gudynas, 2019), and it is based on both ecosystems and communities as part

3 Article 24 of Chile's constitution establishes that the state has *inalienable, imprescriptible rights, eminent domain over the mineral wealth of the country*, which can be privatized by a regime of private concessions. Water rights, too, are regulated by a specific water code (Codigo de Aguas) as national economic goods that can be privatized (Larraín, 2006).

4 A. Acosta (economist and former president of the Ecuador Constitutional Assembly), M. Svampa (sociologist), Y. Herrero (anthropologist and ecofeminist activist) and E. Gudynas (director of the Latin American Center for Social Ecology [CLAES]).

of nature, conceived of as in itself a living entity: Pachamama (Mother Earth)[5] (Acosta, 2019). This conception was integrated into a national constitution for the first time in Ecuador, and even though it has an indigenous matrix, it combines Western deep ecology and indigenous biocentric visions (Acosta, 2019). Pachamama thus constitutes a juridic hybrid and a form of conceptual *mestizaje*, an interpretation of Nature that has a specific territorial, social and cultural space (Acosta et al., 2018). In practice, this vision is intertwined with the counter-hegemonic one of *buen vivir* (living well), which is an evolving and contested concept (Svampa, 2012) nurtured by indigenous cultures[6] and integrated into the Ecuadorian and Bolivian constitutions (Gudynas, 2019; Acosta et al., 2018). This paradigm underlies an alternative economic model, which is opposed to the linear growth of capitalism and based on the balance between the fundamental needs of humanity and the resources available to satisfy them (De la Cuadra, 2015). As a form of decolonization from Western lifestyles and social relationships, *buen vivir* refers to reciprocity, solidarity, complementarity and a decommodification of life and work (Marañón Pimentel et al., 2016). It is linked to the ecofeminist ethics of caring (León 2011; Herrero 2019), involving a strong ecological dimension that directly refers to the conception of Pachamama (Gudynas, 2011).

The relevance of this strand of thought in the context of Latin American climate action is shown by the manifestos of the SCAC and of Cumbre de los Pueblos assemblies, which declared:

5 This term comes from Andean tradition and is commonly used to define an alternative conception of a relationship with nature that is more profound than the anthropocentric relationship. Gudynas clarifies that: "Pachamama refers to the environment in which the person is inserted. The classic European duality that separates society from Nature ... does not apply here. In the Andean world, this distinction does not exist since people are part of the environment, and their idea of environment is not only biological or physical, but it is also social. Simultaneously, this relationship cannot be understood as the interaction of an individual within the Pachamama. Human interactions are always collective, belonging of a community and not of isolated individuals.... The Andean concept of community is also different from the Western one.... [I]t includes non-human living beings, such as certain animals or plants, some non-living elements, in particular hills or mountains, and the spirits of the ancestors. In addition, these communities are related to a certain territory ... as a way of understanding oneself as part of a socially and ecologically expanded community, inserted in an environmental and territorial context" (Gudynas, 2019, p. 108).

6 *Sumak Kawsay* (Quechua), *Suma Qamaña* (Aymara) and *Küme Mongen* (Mapugundun) are indigenous *cosmovisiónes* that go beyond their use in the native population or specific social groups to represent a Latin American cultural diversity that recognizes the plurality of ways of living, values and the aim to live collectively in aware and responsible relationships based on the interdependency of all species and Mother Earth (De la Cuadra, 2015).

> The challenge that the climate crisis imposes on us cannot be restricted exclusively to the reduction of our emissions, but it must involve the construction of a new relationship with Pachamama, focused on the integral respect of all forms of life as a founding principle of our communities, cultures and territories. (SCAC, 2019)

> We invite ourselves to collectively cocreate ways to relate to nature, and a conscious and loving way to relate to each other.... We understand the need to incorporate the rights of nature and of all living beings that sustain the planetary fabric, as constituents of our lives and the laws of the peoples. (Cumbre de los Pueblos, 2019)

The discussions in both assemblies framed climate action as a decolonization of human relationships with Pachamama, where the protection of elements essential for the development of life goes beyond their utility to humanity. In the manifestos of Cumbre de los Pueblos and SCAC, major attention is given to: (1) recognition of the rights of nature, (2) deprivatization of water and protection of the commons, (3) just transition, (4) shifting the development model toward *buen vivir*, (5) gender perspectives and ecofeminism, (6) food and energy sovereignty, (7) participatory governance, (8) recognition of indigenous peoples' and Afro-descendants' demands and (9) respect for human rights and signing the Escazú Agreement.[7]

This radical climate agenda fostered by the Chilean grassroots organizations reflects broader and older Latin American debates and struggles concerning environmental justice. From participant observation to the SCAC and Cumbre de los Pueblos assemblies, the demand for a new constitution captured people's attention and concepts like the rights of nature and *buen vivir* related to specific territorial claims became fundamental.

Urban adaptation and neoliberal urbanism

There is an ongoing process of embedding the grassroots demands described above in the new constitution of Chile. This will affect the urban agenda and adaptation, especially in relation to the neoliberal urbanism that

7 The Escazú Agreement is a legally binding regional Latin America treaty that includes the protection of human rights defenders in environmental matters. After being one of the major actors in its negotiations, Chile refused to sign the Escazú Agreement, proving that its engagement with it was only instrumental—a way to getting the COP25 presidency.

characterizes Chilean urban development (Peck et al., 2009). Indeed, the application of the neoliberal model to planning and territorial governance have generated urban processes dominated by private interests (Carraro et al., 2021; Bustos Gallardo et al., 2019; Maturana, 2017). In recent years, despite some progress in urban adaptation (Barton et al., 2021), the market-driven and technocratic logic of the Chilean planning framework has not been challenged.

In order to understand how the neoliberal urbanism model influences urban adaptation while reproducing socio-ecological inequalities, three points must be considered: (1) liberalization of land use, (2) scarce integration of risk as multidimensional and community claims in planning, and (3) climate inequalities and the commodification of urban nature.

First, the 1979 National Urban Development Policy (PNDU) contributed to creating a planning framework congruent with a neoliberal socioeconomic model (Trivelli, 1981) by establishing the priority of profitability in land use while confining the state to a subsidiary role as provider of infrastructure and facilities (Carraro et al., 2021). Since land is *not* considered a scarce resource, the idea was that the market would pose limits to urban development through prices without the need for urban development limitations (Jimenez et al., 2018). This resulted in a relocation of the poor to peripheral areas lacking access to basic services, infrastructure and job opportunities. Often these areas are more prone to natural hazards, thereby exposing vulnerable groups to socio-ecological disasters (Vásquez et al., 2017; Romero, 2014). On the other hand, neoliberal development contributed to an aggressive urbanization that damaged both landscapes and urban heritage (Maturana, 2017; Hidalgo et al., 2008; Carraro et al., 2021).

The extreme consequences of this land-use liberalization in 1985 led the PNDU to recognize the need for the state's role in regulating land use and in posing limits to urban expansion (Maturana, 2017). The 2013 adjustment to this policy promoted social integration and environmental sustainability, but did not change the main framework based on the commodification of both landscapes and nature and the prevalence of real estate extractivism (Irarrázaval, 2012; Hidalgo et al., 2016). Furthermore, the social housing policy developed by the different governments in the last 40 years has been implemented by private ventures and largely contributed to urban expansion and segregation (Ducci, 1997). Evictions were systematically conducted from 1979 to 1985, dismantling informal and poor settlements in central areas. About 170,000 inhabitants were relocated to peripheral and peri-urban social housing compounds to create new spaces for the expanding neoliberal city (Sanzana Calvet, 2016; Rodriguez & Rodriguez, 2009).

Second, despite Chile's high exposure to geophysical and climate-related natural hazards, the nation's existent planning tools do not include risk as a factor in decision-making (Gonzales et al., 2018; Romero, 2014). Nor are its participatory mechanisms sufficiently integrated to include vulnerable communities (Anguelovski et al., 2016). Together, these deficiencies have been identified as of the major drivers in the reproduction and persistence of the Chilean neoliberal urban development: risk zoning does not prevent real estate expansion from pushing disadvantaged groups to live in hazardous sites (Maturana, 2017; Romero & Mendes, 2020), and the unequal distribution of risks reinforces scenarios of environmental injustice (Sandoval, 2015).

A growing debate in recent years focuses on how risk and resilience cannot be considered mere technical aspects—as they are today—in Chilean planning (Gonzales et al., 2018; Romero, 2014). The influence of the built environment and social dynamics on risk is overlooked, while conflicting interests and the priority of private property dictate policies (Visconti et al., 2021; Sandoval, 2020). Vulnerable groups are systematically ignored as well as the complex relationships between land markets, planning regulations, environmental degradation and disaster vulnerability (Carraro et al., 2021). Community claims and local knowledge are neglected, and the modes and extent of participation in planning vary from place to place, creating opportunities for economic speculation and undermining public trust in the process (Herrmann et al., 2016; Vásquez et al., 2017).

Third, the neoliberal urban development model uses nature as a real-estate "consumption instrument" (Irarrázaval, 2012), generating urbanization patterns that are exacerbating climate inequalities (Romero & Mendes, 2020, 2021). This process has manifested in practices of new urban environmentalism (Sanzana Calvet, 2016). This is particularly evident in Santiago de Chile, where urban greening and green enclaves are a feature of the real estate development to address the "green imaginary" of high-income groups (Irarrázaval, 2012). This commodification of nature in the city, made by an instrumentalization of urban greening and ecological processes to increase the commercial value of housing, generated an unequal distribution of green areas and environmental benefits (Vásquez et al., 2017). The features of this process can be described by three trajectories:

1. Consolidation of the pattern of segregation that was defined during the dictatorship with the localization of high-income groups in the eastern neighborhoods, served by green areas, and with low-income groups occupying peri-central or peripheral areas in the west, south and north with the worst environmental assets (Sanzana Calvet, 2016).

2. Creation of green enclaves in the peri-urban belt, with the urbanization of rural or semi-rural areas as sustainability experiments and allowed by the 1994 metropolitan masterplan, the Plan Regulador Metropolitano de Santiago (PRMS) (Sanzana Calvet & Castán Broto, 2016).
3. Reinforcement of these socio-spatial inequalities: the systematic privatization of infrastructures and services (transport/energy, water) (Sanzana Calvet & Castán Broto, 2015); access to health, education and employment (CASEN, 2020; Méndez & Otero, 2018); social housing policies that relocate poor people to peripheral areas (Rodriguez et al., 2014); and evictions of informal settlements from valuable sites.

These patterns of segregation affect the low-income urban neighborhoods most, highlighting an uneven distribution of environmental conditions and allocation of economic and social assets. This creates climate inequalities within the urban fabric (Irarrázaval, 2012; Romero & Mendes, 2021). From the detailed studies on climate inequalities (Romero & Mendes, 2020, 2021) and on the distribution of environmental beneficial conditions and housing value (Irarrázaval, 2012), indeed the city appears separated in two blocks: (1) the rich eastern neighborhoods with a high rate of green areas, less air pollution and less urban heat island effect, and (2) the western neighborhoods with thermal discomfort, lack of ventilation and air pollution.

The urban adaptation initiatives[8] developed in recent years also retain this logic. Central high-income and medium-income districts are consolidating a network of green infrastructures (such as parks, river waterfronts and green bicycle lanes) or eco-friendly initiatives (e.g., electric transportation) that are essentially increasing their commercial value. On the contrary, low-income neighborhoods are experiencing water shortages and difficulties with maintaining green areas, and are threatened by real estate speculation, criminality and reduced maintenance.

Ecobarrios: Coproducing resilience practices

Despite being a crucial issue for climate action in Chile, the urban agenda disappeared from the main debate during the Cumbre de los Pueblos,

8 These include Climate Adaptation Santiago (CAS) in 2014, the creation of the Municipal Network for Climate Change by the NGO Adapt Chile, the Santiago Metropolitan Region Strategy, Santiago Humano y Resiliente of 2017, funded by the Rockefeller Foundation, and the creation of the Regional Climate Change Councils (CORECCs) in 2018.

which was oriented toward territorial socio-environmental conflicts and the call for a new constitution. Within the activities of the Cumbre de los Pueblos, a space was dedicated to reflection and collective spirituality (Aldea de la Paz). Here, discussions and workshops on practices of *buen vivir* and transition took place, led by the Chilean Institute of Permaculture and Transition.[9] In Chile, activities related to the Transition movement started in 2010 and have progressively been interlinked with the practices of the *ecobarrios*, born during the last decade in different neighborhoods of Santiago de Chile.

Ecobarrios are bottom-up initiatives that began in the vulnerable neighborhood of Maipù. They are conceived as:

> self-organized communities whose aim is to improve the communities' quality of life as well as taking care of the environment. Their pillars are: human relations, ecological education and the creation of networks of organizations and people committed to changing the current model. It must be clarified that it is not a sum of places and a set of alternative solutions for sustainable energy or for recycling. (Ecobarrio Maipù statement)

These bottom-up actions question the mainstream acceptance of "eco-neighborhoods," creating self-organized practices that redraw the concept of green neighborhoods through a transformative interpretation. The green and top-down neoliberal logic behind the eco-neighborhood rhetoric is subverted toward a community-centered and grassroots form of urban adaptation (Schlosberg & Collins, 2014). The practices developed in the *ecobarrios* have a strong focus not only on building self-sufficiency through community gardening, recycling and water storage, but also on coproduction, here intended as a shift in power relationship toward participatory services and knowledge creation (Petrescu et al., 2016). A distinctive feature, in fact, is that workshops, courses and activities are open to everyone and based on the criterion of reciprocity. Micro-scale actions and engagement in macro-scale socio-environmental topics are both pursued. The aim is to educate everyone who is willing to participate in and discuss socio-ecological and political issues about sustainable ways of living and practices (such as gardening,

9 The Transition movement is a civil society movement that was born in UK (2005) with the aim of leading local, self-organizing actions capable of responding to climate change through transition toward low-carbon lifestyles, self-sufficiency, community empowerment and participatory governance with a focus on food, energy and education (Seyfang, 2012; Hopkins, 2011).

composting, phyto-cosmetics and water recycling). Participation is highly heterogeneous, involving people with diverse backgrounds in terms of age, education, and political and social engagement. These actions foster an empowerment process that resists systemic urban inequalities while generating awareness about environmental justice issues.

The everyday practices include local food production and gardening, recycling and self-production of products for daily life. Organized barter and redistribution of surpluses are also incentivized through time banking to exchange needs and skills and share small domestic appliances and secondhand clothes. Educational activities about sustainability for both seniors and youth are developed on a weekly basis and offered by volunteers. In this everyday dimension of mutual support, *ecobarrios* are places of resistance to neoliberal urbanism and real estate development. They reshape the fabric of the city by offering places for community action, exchange of knowledge and tangible support free from market dynamics. *Ecobarrios* participants cocreate community hubs that subvert the scarcity of economic resources and services that characterizes the segregated dynamics of Santiago. They provide an active network of daily action based on community assets.

Insights from fieldwork and in-depth interviews showed that *ecobarrios* can be considered a radical form of adaptation (Dawson, 2017) developed by the more marginalized communities in Santiago against the rise of climate inequalities. For instance, the *ecobarrio* of Maipù originated from a conflict with the municipality responsible for destroying the Parque 4 Alamos by cutting down 119 trees. The Yungay and Villa Frei *ecobarrios* developed to protect historical heritage by saving parts of their neighborhoods from real estate speculation. The *ecobarrio* of Villa Portales was developed in one of Santiago's historical sociopolitical resistance enclaves (see Figure 11.2).

Ecobarrios generally occur in or near social housing blocks built in the 1960s and 1970s. Thus the *ecobarrio* experience is rooted in a sense of community historically engaged in resisting the political and socioeconomical oppression of dictatorship. This sociopolitical identity was particularly evident during the 18-O upsurge, when some of the *ecobarrios* were centers of political discussion and solidarity action responding to the critical situation generated by the looting of supermarkets, the interruption of public transportation and indiscriminate police violence in Santiago's most marginalized neighborhoods. They became hot spots where food and daily needs were collected for the poorest inhabitants, and assistance was provided to the victims of police violence.

Figure 11.2. Villa Portales, murals made after 18-O, near the *ecobarrio* entrance. Source: Cristina Visconti.

According to the definitions offered by Petrescu et al. (2016) and Baibarac et al. (2017), *ecobarrios* can be considered coproduced resilience practices that substantially shift the business-as-usual resilience discourse. They offer a counter-vision to the market-based and top-down approaches to climate change adaptation that tend to normalize neoliberal ideology and use community empowerment as a strategy to dismantle the public provision of services and the protection of the commons (Cretney & Bond, 2014). In this sense, the grassroots practices of *ecobarrios* question the institutional discourse on urban adaptation by rearticulating resilience as the cocreation of a process of "commoning." This term refers to the social process that creates and reproduces the commons (An Architektur, 2010), and it is used by Petrescu et al. (2016) to suggest that community self-organized adaptation practices are a project of resistance to privatization and globalization (Brown et al., 2012; Armiero, 2021).

Pueblos en Transición, the rights of nature and *buen vivir*

An interesting process emerged during the Cumbre de los Pueblos assembly after an increasing demand from the audience to integrate the transition discourse within the context of social upsurge. In responding to this invitation, the exponents of the *ecobarrios* jointly with the Chilean pioneering transition trainers decided to develop a workshop series, Pueblos en Transición (People in transition), which was held in December 2019 and January 2020. Cumbre de Los Pueblos and SCAC articulated concepts such as *buen vivir*, the rights of nature and post-extractivism transition (e.g., energy and food sovereignty, and participatory governance), making them visible in the Chilean context where they help to build an alternative imaginary in opposition to neoliberal discourse.

The initiative of Pueblos en Transición offers space for reflection about the fact that transition discourse does not correspond to the Latin America socio-ecological perspective. Although informed by the Transition movement tools, those community-led actions seem more inclined toward operationalizing *buen vivir* in practice. Applying *buen vivir* principles like reciprocity, social cooperation and sociocracy to accomplish ecologically sustainable transformations, community-based adaptation practices like the *ecobarrios* are articulating an alternative imaginary to market-driven urban policies. Although an alignment between claims for a post-extractive macro-scale shift and micro-scale actions can be noted, several limitations and gaps can be formulated about how bottom-up processes can inform the Chilean urban adaptation agenda.

Climate justice and the right to the city are not strong components in the *buen vivir* paradigm and in the rights of nature legal frameworks (Acosta et al., 2018). Postcolonial theorizations of urban change are still unrepresented in the debate on climate urbanism (Robin & Castán Broto, 2021). The concepts of *buen vivir* and the rights of nature and the alternative experiences that they inspire can nurture a unique space of reflection about just adaptation.

The recognition of the rights of nature, with their biocentric focus, poses the question of how nature-based processes of adaptation can remain unmoved by the utilitarian value often pursued through the application of the ecosystem services framework (Pagiola et al., 2003). This implies the need to overhaul the relation of urban development with urban nature and its management through the lenses of biocentrism and commoning. This perspective can leverage the application of ecological justice and equity to urban greening, thus explicitly contesting green urbanism (Gram et al., 2021) and the commodification of nature that generates the uneven distribution of environmental benefits in the Chilean urban context.

At the same time the *buen vivir* framework can foster the inclusion of new social, economic and environmental values while offering communities the possibility of "recommoning" the assets necessary to sustain collective activities in their neighborhoods and beyond (Brown et al., 2012). The way the rights of nature and *buen vivir* will be included in Chile's new constitution opens new frontiers in how the planning system can integrate transformative relationships to nature and what forms of knowledge and power need to be transformed toward inclusivity and participation. It will be crucial to promote a situated climate urbanism (Robin & Castán Broto, 2021) that includes the ideas of resilience and adaptation not in terms of adjustment to strengthen local coping capacities, but in terms of transformative perspectives (Sandoval et al., 2020).

These perspectives have to be articulated to reduce socio-territorial vulnerability to embedded social demands (Robin & Castán Broto, 2021) while fostering resolutions of socio-environmental conflicts. Grassroots resilience initiatives like the *ecobarrios* have great potential in informing this shift, but they can fail to acquire strategic capacities to transform the systems in which they operate due to the scarce resources they can mobilize (Petrescu et al., 2016). In this sense, there is a need to overcome the "local trap" (Purcell, 2006), that is, the risk of fostering socially just and ecologically sustainable decision-making at the community level, while remaining disconnected to broader political agendas (Baibarac et al., 2017). In the case of Chile, the opportunity of the constitutional convention and the participation of many socio-environmental activists with clear knowledge of the structural inequalities of the Chilean system can hopefully create a shift in the nation's legal framework. This will be a fundamental step toward alternative governance of both the commons and nature that implies the agency of empowerment and governance by the community, potentially leading to an overhaul of the neoliberal planning system and territorial governance. This political trajectory joined with the grassroots actions here explored will be also a reference for the implementation of urban resilience practices nurtured by nonutilitarian logics and oriented toward reshaping inequalities by combining climate change action with demands for social justice.

Postscript

The chapter was finalized in April 2022 and reports events related to the constitutional process until the beginning of the constitutional convention (October 18, 2021). After one year a draft of the new constitution was developed to be subjected to a referendum on September 4, 2022 (the *plebiscito constitucional de Chile* of 2022). The majority of the voters rejected the draft constitution with only 38% of the votes in favor of its approval. The proposed new constitution dedicated one chapter to environmental matters with important advancements and innovations (Huneeus, 2022; Jimenez et al., 2022). Several articles state principles of environmental justice, the responsibility of the state to respond to the climate crisis, the rights of nature and the category of natural commons as goods not subject to ownership, including water (Convención Constitucional, 2022, Chapter 3, pp. 45–51). Unfortunately, the lack of clear rules to set a new constitutional process opens great uncertainty on how to proceed in changing Chile's old constitution, how the new text will be written and which advancements will be included.

Political analysts and President Gabriel Boric, who leads a left-wing coalition, have argued that the results of the 2020 plebiscite for a new constitution clearly indicated that the majority of Chileans are in favor of reforming Pinochet's constitution (Carrere, 2022). Street protests led by students spread after September 4, 2022, the conflict between the Chilean state and the Mapuche people increased in violence and repression in the south of the country, the right-wing coalition demands to be more included in the rewriting of the new constitution, and a severe economic crisis is affecting low- and middle-income groups already threatened by the socioeconomic consequences of COVID-19 and inflation (Kauffmann, 2022; CEPAL, 2022). Given this national context and considering the wider global context, the directions that the struggle for a new constitution will take are highly unpredictable.

References

Acosta, A. (2019). *La naturaleza como subjecto de derechos: La gran tarea pendiente.* Cumbre de los Pueblos Santiago de Chile, December 2019.

Acosta, A., & Brand, U. (2018). *Salidas del laberinto capitalista: decrecimiento y postextractivismo.* OLCA.

Allain, M. (2019). Conflictos y protestas socio-ambientales en Chile: Reflexiones metodológicas y resultados. *Revista de Sociología, 34*(1), 81–101. https://doi.org/10.5354/0719-529X.2019.54271

An Architektur. (2010). On the Commons: A Public Interview with Massimo De Angelis and Stavros Stavrides, *E-flux, 17*. https://www.e-flux.com/journal/17/67351/on-the-commons-a-public-interview-with-massimo-de-angelis-and-stavros-stavrides/

Anguelovski, I., Shi, L., Chu, E., Gallagher, D., Goh, K., Lamb, Z., Reeve, K., & Teicher, H. (2016). Equity impacts of urban land use planning for climate adaptation: Critical perspectives from the global north and south. *Journal of Planning Education and Research, 36*(3), 333-348.

Armiero, M. (2021). *Wasteocene: Stories from the global dump*. Cambridge University Press.

Azócar, G., Romero, H., Sanhueza, R., Vega, C., Aguayo, M., & Muñoz, M. D. (2007). Urbanization patterns and their impacts on social restructuring of urban space in Chilean mid-cities: The case of Los Angeles, Central Chile. *Land Use Policy, 24*(1), 199–211. https://doi.org/10.1016/j.landusepol.2005.04.003

Baibarac, C., & Petrescu, D. (2017). Co-design and urban resilience: Visioning tools for commoning resilience practices. *CoDesign, 15*(2), 91–109. https://doi.org/10.1080/15710882.2017.1399145

Barkin, D., & Lemus, B. (2016). Local solutions for environmental justice. In F. de Castro, B. Hogenboom, & M. Baud (Eds.), *Environmental governance in Latin America* (pp. 257–286). Palgrave Macmillan.

Barton, J. R., Gutiérrez-Antinopai, F., & Escalona Ulloa, M. (2021). Adaptive capacity as local sustainable development: Contextualizing and comparing risks and resilience in two Chilean regions. *Sustainability, 13*(9), 4660. https://doi.org/10.3390/su13094660

Bollier, D. (2008). Los bienes comunes: un sector soslayado de la riqueza. In H. Silke (Ed.), *Genes, bytes y emisiones: Bienes comunes y ciudadanía*. Fundación Heinrich Böll.

Borsdorf, A., Hidalgo, R., & Sánchez, R. (2007). A new model of urban development in Latin America: The gated communities and fenced cities in the metropolitan areas of Santiago de Chile and Valparaíso. *Cities, 24*, 365–378. https://doi.org/10.1016/j.cities.2007.04.002

Bouyé, M. (2019, November). Chile's protests offer lessons on social inequality and climate action. World Resources Institute. https://www.wri.org/insights/chiles-protests-offer-lessons-social-inequality-and-climate-action

Brown, G., Kraftl, P., & Pickerill, J. (2012). Holding the future together: Towards a theorisation of the spaces and times of transition. *Environment and Planning A, 44*(7), 1607–1623. https://doi.org/10.1068/a44608

Bustos Gallardo, B., Lukas, M., Stamm, C., & Torre, A. (2019). Neoliberalismo y gobernanza territorial: propuestas y reflexiones a partir del caso de Chile, *Revista de geografía Norte Grande, 73*, 161–183. https://doi.org/10.4067/S0718-34022019000200161

Carraro, V., Visconti, C., & Inzunza, S. (2021). Neoliberal urbanism and disaster vulnerability on the Chilean central coast. *Geoforum, 121*(3), 83–92. https://doi.org/10.1016/j.geoforum.2021.02.023

Carrere, M. (2022, September 16). Chile: ¿Qué viene para el medio ambiente tras el rechazo de la propuesta de constitución? Mongabay. https://es.mongabay.com/2022/09/que-viene-para-el-medio-ambiente-tras-el-rechazo-de-la-propuesta-de-constitucion-en-chile/

CASEN. (2020). Encuesta de carcacterización socioeconómica nacional. Ministerio de Planificación. Santiago de Chile.

Castán Broto, V., & Sanzana Calvet, M. (2020). Sacrifice zones and the construction of urban energy landscapes in Concepción, Chile. *Journal of Political Ecology, 27*(1), 279–299. https://doi.org/10.2458/v27i1.23059

CEPAL. (2012). *La economía del cambio climático en Chile*. Comisión Económica para América Latina y el Caribe.

CEPAL. (2022). *Balance preliminar de las economías de América Latina y el Caribe*. Comisión Económica para América Latina y el Caribe.

Convencion Constitutional. (2022). Propuesta Constitucion Politica de la Republica de Chile. https://www.chileconvencion.cl/wp-content/uploads/2022/07/Texto-Definitivo-CPR-2022-Tapas.pdf

Correa-Parra, J., Vergara-Perucich, J. F., & Aguirre-Nuñez, C. (2020). Water privatization and inequality: Gini coefficient for water resources in Chile. *Water, 12*(12), 3369. https://doi.org/10.3390/w12123369

Cretney, R., & Bond, S. (2014). 'Bouncing back'to capitalism? Grass-roots autonomous activism in shaping discourses of resilience and transformation following disaster. *Resilience*, 2(1), 18-31.

Cumbre de los Pueblos. (2019). *Declaración Cumbre de los Pueblos*. Santiago de Chile. https://cumbresocialclima.net/declaracion-cumbre-pueblos/

Dawson, A. (2017). *Extreme cities: The peril and promise of urban life in the age of climate change*. Verso Books.

De la Cuadra, F. (2015). Buen vivir ¿Una auténtica alternativa post-capitalista? *Polis, 40*. http://journals.openedition.org/polis/10893

Ducci, M. E. (1997). Chile: el lado oscuro de una política de vivienda exitosa. *EURE, 23*(69), 99–115. http://www.eure.cl/index.php/eure/article/view/1164

ECLAC. (2018). Social Panorama of Latin America, LC/PUB.2019/3-P, Santiago, 2019.

Elacqua, G. (2012). The impact of school choice and public policy on segregation: Evidence from Chile. *International Journal Educational Development, 32*(3), 444–453. https://doi.org/10.1016/j.ijedudev.2011.08.003

Esteva, G. (2014). Commoning in the new society. *Community Development Journal, 49*(Suppl. 1), i144-i159. https://doi.org/10.1093/cdj/bsu016

Fernández, I., Manuel-Navarrete, D., & Torres-Salinas, R. (2016). Breaking resilient patterns of inequality in Santiago de Chile: Challenges to navigate towards a more sustainable city. *Sustainability, 8*(8), 820. https://doi.org/10.3390/su8080820

Global Carbon Project. (2018). *Global carbon atlas*. http://www.globalcarbonatlas.org/en/CO2-emissions

Gonzalez D., Monsalve M., Moris, R., & Herrera, C. (2018). Risk and resilience monitor: Development of multiscale and multilevel indicators for disaster risk management for the communes and urban areas of Chile. *Applied Geography, 94*, 262–271. https://doi.org/10.1016/j.apgeog.2018.03.004

Gram, Y., Fourie, E., & Marshall, S. (2021). *Exploring urban rights for nature: Possibility, praxis + pathways*. School of Community and Regional Planning. University of British Columbia, PLAN 526. https://doi.org/10.14288/1.0398223

Gudynas, E. (2011). Tensiones, contradicciones y oportunidades de la dimensión ambiental del Vivir Bien. In I. Farah & L. Vasapollo (Eds.), *Vivir bien: ¿Paradigma no capitalista?* (pp. 231–246). CIDES-UMSA y Plural.

Gudynas, E. (2019). *Derechos naturaleza: Ética biocéntrica y políticas ambientales*. Editorial Quimantu.

Gudynas, E. (2020). Derechos de la naturaleza en Chile constituyente, conversaciones eco-constituyentes: un programa sobre desafíos socioambientales. OLCA. http://accionyreaccion.com/derechos-de-la-naturaleza-en-chile-constituyente/

Gudynas, E. (2021, March). Propiedad, acceso y excedentes de los recursos naturales ante la constituyente de Chile. *CLAES*. http://extractivismo.com/2021/03/propiedad-acceso-y-excedentes-de-los-recursos-naturales-ante-la-constituyente-de-chile/

Harvey, D. (2003). Le Nouvel impérialisme: accumulation par expropriation. *Actuel Marx, 35*, 71–90. https://doi.org/10.3917/amx.035.0071

Hidalgo, R., Borsdorf, A., Zunino, H., & Álvarez, L. (2008). Tipologías de expansion metropolitana en Santiago de Chile: precariópolis estatal y privatópolis inmobiliaria. In *Diez años de cambios en el Mundo, en la Geografía y en las Ciencias Sociales, 1999-2008. Actas del X Coloquio Internacional de Geocrítica*, Universidad de Barcelona, 26–30 de mayo de 2008. http://www.ub.es/geocrit/-xcol/434.htm

Hidalgo, R., Camus, P., & Paulsen Espinoza, A. (2016). Extractivismo inmobiliario, expoliacion de los bienes comunes y esquilmacion del medio natural. El borde costero en la macrozona central de Chile en las postrimerías del neoliberalismo. In M. Coy, J. Stötter, & A. Borsdorf (Eds.), *Die Welt verstehen—eine geographische Herausforderung. Eine Festschrift der Geographie Innsbruck für Axel Borsdorf* (pp. 251–270). Innsbrucker Studienkreis für Geographie, Universität Innsbruck.

Herrero, Y. (2019). *Ecofeminismos para el buen vivir: vivir y trabajar en un mundo justo y sostenible*. Ideas Commercio Justo, Barrios por el comercio Justo, Ayuntamiento de Córdoba.

Herrmann, M. G., & Van Klaveren, A. (2016). Disminución de la participación de la población en organizaciones sociales durante los últimos trece años en Chile e implicaciones para la construcción de una política de planificación

urbana más participativa *EURE, 42*(125), 175–203. https://doi.org/10.4067/S0250-71612016000100008
Holst, J., Laaser, U., & Hohmann, J. (2004). Chilean health insurance system: A source of inequity and selective social insecurity. *Journal of Public Health, 12*(4), 271–282. https://doi.org/10.1007/s10389-004-0026-5
Hopkins, R. (2011). *The transition companion*. Green Books.
Huneeus, A. (2022, August 31). Win or lose, Chile's draft constitution heralds a new era of climate constitutionalism. *VerfBlog*. https://verfassungsblog.de/win-or-lose/. https://doi.org/10.17176/20220831-181822-0
INDH. (2021). *Mapa de conflictos socio-ambientales de Chile*. https://mapaconflictos.indh.cl/#/
Irarrázaval, F. (2012). El imaginario "verde" y el verde urbano como instrumento de consumo inmobiliario: configurando las condiciones ambientales del área metropolitana de Santiago. *Revista INVI, 27*(75), 73–103. https://doi.org/10.4067/S0718-83582012000200003
Jiménez, G., Astorga, M., Cuadra, V., Bernales, F., & Avendaño, L. (2022, June). Análisis Propuesta Constitucional N22, Medio-Ambiente. Cariola Diez Peres-Cotapo. https://www.cariola.cl/en/actualidad-constitucional/22-medio-ambiente/
Jimenez, V., Hidalgo, R., & Campesino, A.-J. (2018). Normalización del modelo neoliberal de expansión residencial mas allá del límite urbano en Chile y España. *EURE, 44*(132), 17–46. http://www.eure.cl/index.php/eure/article/view/2379
Kaufmann, R. (2022, September 8). Chile's constitutional endeavour goes on *VerfBlog*. https://verfassungsblog.de/chile-goes-on/. https://doi.org/10.17176/20220908-230743-0
Kreft, S., Eckstein, D., & Melchior, I. (2017). *Global climate risk index 2017: Who suffers most from extreme weather events? Weather-related loss events in 2015 and 1996 to 2015*. Germanwatch e.V.
Larraín, S. (2006). El agua en Chile: entre los derechos humanos y las reglas del mercado. *POLIS: Revista Latinoamericana, 14*. https://www.redalyc.org/articulo.oa?id=30551406
León M. (2011). Sumak Kawsay y feminismo: las mujeres en la construcción del Buen Vivir. *Südwind Magazine, 2*.
Machado Aráoz, H. (2012, November). Los dolores de *Nuestra América* y la condición neocolonial. Extractivismo y biopolítica de la expropiación. *OSAL, 13*(32).
Machado Aráoz, H. (2009). Auge minero y dominación neocolonial en América Latina. Ecología política de las transformaciones socioterritoriales neoliberales. Paper presented at XXVII Congreso de la Asociación Latinoamericana de Sociología. Asociación Latinoamericana de Sociología, Buenos Aires.
Marañón Pimentel, B., & López Córdoba, D. (2016). Del desarrollo capitalista al Buen Vivir desde la descolonialidad del poder. *Intersticios de la política y la*

cultura. Intervenciones Latinoamericanas, 5(10), 5–20. https://revistas.unc.edu.ar/index.php/intersticios/article/view/15767

Marconi, A. (2021, June 6). *Constitución ¿Ecológica?* Heinrich Böll Stiftung. https://cl.boell.org/es/2021/07/06/constitucion-ecologica

Martínez, C. (2015). El análisis y la gestión de riesgos naturales en ciudades intermedias y localidades pequeñas en Chile. In F. Maturana & A. Rojas (Eds.), *Ciudades intermedias en Chile: Territorios olvidados* (pp. 105–125). RIL Editores.

Maturana, F. M. (2017). ¿Ausencia de planificación urbana en Chile? Algunas reflexiones. *Cybergeo: European Journal of Geography.* http://journals.openedition.org/cybergeo/28064

Méndez, M. L., & Otero, G. (2018). Neighbourhood conflicts, socio-spatial inequalities, and residential stigmatisation in Santiago, Chile. *Cities, 74*, 7–82. https://doi.org/10.1016/j.cities.2017.11.005

Navarro, F., & Tromben, C. (2019). "Estamos en guerra contra un enemigo poderoso, implacable": los discursos de Sebastián Piñera y la revuelta popular en Chile. *Literatura y lingüística, 40*(40), 295–324. https://doi.org/10.29344/0717621x.40.2083

OECD. (2021). Income inequality (indicator). https://doi.org/10.1787/459aa7f1-en

Oyarzún Serrano, L. (2021). *Independientes, pueblos indígenas y mujeres en la nueva Constitución de Chile¿fin de un modelo?* Fundación Carolina. https://doi.org/10.33960/AC_18.2021

Pagiola, S., Callicott, J. B., De Groot, R. S., & Hassan, R. (2003). Concepts of ecosystem value and valuation approaches. In World Resources Institute (Ed.), *Ecosystems and human well-being: A framework for assessment* (pp. 127–147). Island Press.

Peck, J., Theodore, N., & Brenner, N. (2009). Neoliberal urbanism: Models, moments, mutations, *SAIS Review of International Affairs, 29*(1), 49–66.

Petrescu, D., Petcou C., & Baibarac, C. (2016). Co-producing commons-based resilience: Lessons from R-Urban. *Building Research & Information, 44*(7), 717–736. https://doi.org/10.1080/09613218.2016.1214891

PNUD. (2018). *Evolución de la pobreza 1990–2017: ¿Cómo ha cambiado Chile?* PNUD.

Politzer, P. (2021, July 25). El extraordinario tintineo en los pasillos. *The Clinic.* https://www.theclinic.cl/2021/07/25/columna-de-patricia-politzer-el-extraordinario-tintineo-en-los-pasillos/?fbclid=IwAR0RUQrAk9pCYFPufSlMGp6PNjH4yapesMzmXqRI13UkEvuWoB3SvzJXep4

Poo, P. (2021, May 20). *Análisis del resultado de las elecciones a la convención constitucional: un abordaje desde los bienes de la naturaleza.* Heinrich Böll Stiftung. https://cl.boell.org/es/2021/05/20/analisis-del-resultado-de-las-elecciones-la-convencion-constitucional

Posner, P. W. (2012). Targeted assistance and social capital: Housing policy in Chile's neoliberal democracy. *International Journal of Urban and Regional Research, 36*(1), 49–70. https://doi.org/10.1111/j.1468-2427.2011.01059.x

Purcell, M. (2006). Urban democracy and the local trap. *Urban Studies, 43*(11), 1921–1941. https://doi.org/10.1080/004209806008978

Robin, E., & Castán Broto, V. (2021). Towards a postcolonial perspective on climate urbanism. *International Journal of Urban and Regional Research, 45*(5), 869–878. https://doi.org/10.1111/1468-2427.12981

Rodríguez, A., & Rodríguez, P. (Eds.). (2009). *Santiago: Una ciudad neoliberal.* Olacchi.

Rodriguez, A., Rodriguez, P., Saborido, M., Segovia, O., & Mires, L. (2014). Visible and invisible violence and inequality in neoliberal Santiago. *Environment and Urbanization, 26*(2), 359–372. https://doi.org/10.1177/0956247814542723

Romero, H. (2014). Vulnerabilidad, resiliencia y ordenamiento territorial de los desastres socionaturales en Chile. *Polígonos. Revista de geografía, 26*, 87–110. https://doi.org/10.18002/pol.v0i26.1700

Romero, H., & Mendes, F. H. (2021). La construcción social de climas urbanos y su relación con la pandemia de Covid-19 en Santiago de Chile. *Cuadernos de geografía: Revista colombiana de geografía*, 30(2), 376–395. https://doi.org/10.15446/rcdg.v30n2.88701

Romero, H., & Mendes, F. H. (2020). Climatología geográfica crítica de Santiago de Chile. In M. de Oliveira Moura, D. Becerra Lucena, C. Cunico, & C. M. da Silva Moura (Eds.), *Climatologia geográfica: do local ao regional e dimensões socioambientais* (pp. 21–43). Editora UFPB.

Sandoval, J. (2020). Vulnerabilidad-resiliencia ante el proceso de riesgo-desastre: Un análisis desde la ecología política. *Polis, 19*(56), 138–154. http://dx.doi.org/10.32735/s0718-6568/2020-n56-1527

Sandoval, V., Boano, C., González-Muzzio, C., & Albornoz, C. (2015). Explorando potenciales vínculos entre resilencia y justicia ambiental: el caso de Chaitén, Chile. *Magallania (Punta Arenas), 43*(3), 37–49. https://doi.org/10.4067/S0718-22442015000300004

Sandoval, V., Wisner, B., & Voss, M. (2021). Natural hazards governance in Chile. *Oxford research encyclopedia of natural hazard science.* https://oxfordre.com/naturalhazardscience/view/10.1093/acrefore/9780199389407.001.0001/acrefore-9780199389407-e-364

Santibáñez, F. (2016). *El cambio climático y los recursos hídricos de Chile La transición hacia la gestión del agua en los nuevos escenarios climáticos de Chile.* Ministerio de Agricultura Oficina de Estudios y Políticas Agrarias.

Sanzana Calvet, M. (2016). *The greening of neoliberal urbanism in Santiago de Chile: Urbanisation by green enclaves and the production of a new socio-nature in Chicureo.* PhD thesis, University College London.

Sanzana Calvet, M., & Castán Broto, V. (2015). Neoliberal shock, infrastructure disruption, and restructuring in Chile. *Critical Planning, 22*, 11–33.

Sanzana Calvet, M., & Castán Broto, V. (2016). green enclaves, neoliberalism and the constitution of the experimental city in Santiago de Chile. In J. C. Evans, A. Karvonen, & R. Raven (Eds.), *The experimental city* (pp. 107–121). Routledge.

Schlosberg, D., & Collins, L. B. (2014). From environmental to climate justice: Climate change and the discourse of environmental justice. *WIREs Climate Change, 5*(3), 359–374. https://doi.org/10.1002/wcc.275

Seyfang, G., & Haxeltine, A. (2012). Growing grassroots innovations: Exploring the role of community-based initiatives in governing sustainable energy transitions. *Environment and Planning C: Government and Policy, 30*(3), 381–400. https://doi.org/10.1068/c10222

Smart Larrain, J. S. (2019). *Local resistance to extractivism: Community mobilisation in the case of Chile*. PhD diss., University College London.

Sociedad Civil por la Acción Climática. (2019). *Manifesto Latin americano por el clima*. Santiago de Chile. https://www.porlaaccionclimatica.cl/wp-content/uploads/2019/12/Manifiesto-Latinoamericano-por-el-Clima.pdf

Solís, C., & Velásquez, F. (2021, May 19). Chile verde: El ascenso político de los ambientalistas en la Convención y los territorios. *Inteferencia*. https://interferencia.cl/articulos/chile-verde-el-ascenso-politico-de-los-ambientalistas-en-la-convencion-y-los-territorios

Svampa, M. (2012). Consenso de los commodities, giro ecoterritorial y pensamiento crítico en América Latina. *Revista OSAL, Observatorio Social de América Latina, 13*(32) (special issue on Movimientos socioambientales en América Latina edited by E. Sader & P. Gentili), 15–38.

Trivelli, P. (1981). Reflexiones en torno a la Política Nacional de Desarrollo Urbano. *Revista EURE-Revista de Estudios Urbano Regionales, 8*(22), 43–64.

Valenzuela-Fuentes, K., Alarcón-Barrueto, E., & Torres-Salinas, R. (2021). From resistance to creation: Socio-environmental activism in Chile's "sacrifice zones." *Sustainability*, 13(6), 3481. https://doi.org/10.3390/su13063481

Vásquez, A., Lukas, M., Salgado M., & Mayorga, J. (2017). Urban environmental (in) justice in Latin America. In R. Holifield, J. Chakraborty, & G. Walker (Eds.), *The Routledge handbook of environmental justice* (pp. 556–566). Routledge.

Visconti, C., Carraro, V., & Inzuza, S. (2021). *Mapeo colectivo como metodología participativa para el estudio de riesgos de la planificación urbana comunal: una propuesta para Chile*. CIGIDEN, Policy Paper Series. https://www.cigiden.cl/wp-content/uploads/2021/12/PP_MapeoColectivo_ISBN_DIGITAL-1.pdf

World Bank. (2019). *Global economic prospects, June 2019: Heightened tensions, subdued investment*. World Bank Group.

About the author

Cristina Visconti is an architect with a PhD in architectural technology and environmental design. After a period as postdoctoral researcher at CIGIDEN (Centro de Investigación para la Gestión Integrada de Desastres Naturales) and IUET (Instituto de Estudios Urbanos y Territoriales) at the Pontificia Universidad Católica de Santiago de Chile, she has now joined the School of Architecture of the University of Naples Federico II, in Italy. She has conducted several action-research projects in Italy, Germany and Chile on community based-adaptation.

About the author

12. A user manual for just cities?

Aurash Khawarzad

Abstract

User manuals have changed the nature of knowledge production and the material production of goods and services for many industries and professions. Could a user manual do the same for urban planning and development? In his exploratory chapter, Khawarzad offers an answer to this question by placing the idea of a user manual in the context of radical pedagogy and collaborative urbanism. By extending the concept of do-it-yourself (DIY) urbanism and the right to repair the city, this chapter walks the talk in suggesting concrete tools, research methods, meeting practices, deliverables, implementation tips and strategies for tactical urbanism to take back the power and place it in people's hands to remake the city not as a side project but as an integral part of our everyday lives.

Keywords: DIY urbanism, right to repair the city, tactical urbanism, user manual, radical pedagogy

> *Many Gentlemen in this Nation of good Rank and high Quality are conversant in Handy-Works.... How pleasant and healthy this their Diversion is, their Minds and Bodyes find.*
>
> —*Joseph Moxon*

> *Individual commitment to a group effort—that is what makes a team work, a company work, a society work, a civilization work.*
>
> —*Vince Lombardi*

This chapter considers two main questions: First, what is the potential for creating tools that facilitate self-directed (individual and collective) action in creating and implementing comprehensive, actionable, community-based

---

Armiero, M., E. Turhan & S. Paolo De Rosa (eds), *Urban Movements and Climate Change: Loss, Damage and Radical Adaptation*. Taylor & Francis Group 2023
DOI: 10.5117/9789463726665_CH12

urban planning documents that are focused on justice and sustainability? And, second, what could be the key aesthetic qualities and organizational structures of those tools? In order to answer these questions, this chapter will make connections between two disparate yet related disciplines, one being the production of the user manual, and the other being justice and sustainability in urbanized spaces. The terms "user manual," "instructional manual," "do-it-yourself (DIY)" and "guide," will be used interchangeably in this chapter.

Exploring the concept of a user manual for just and sustainable cities can help us ask deeper questions about systems of urban planning and development, particularly how communities can collaboratively envision *and* implement their own comprehensive plans. In its many iterations, the user manual can serve as a vehicle for helping us ruminate on the potential for combining radical pedagogies, in this case, novel tools for participatory research, and the political process of organizing land-based social movements. User manuals, as we will see, have changed the nature of knowledge production and the material production of goods and services for many industries and professions. Could a user manual do the same for urban planning and development?

The theoretical basis for the existence of a user manual for just cities is the concept of people's planning. People's planning is the belief that community members should be deeply involved in creating and implementing the comprehensive vision for their own urban scenery. Direct and sustained involvement in shaping urban planning is key for building systems (and cities) that affirm local indigenous social and environmental practices, which in turn is necessary for environmental sustainability and social justice—as opposed to cities based on extraction of natural resources and exploitation of human value for the means of enriching the upper class. More about people's planning is written below.

As a document that explores the role of instructional manuals and city-making, this chapter also acts as its own instruction manual of sorts. Each section will offer "next steps" in the form of questions for collective or individual reflection, along with various action items that can assist the reader with beginning the process of creating their own user manual.

The dawn of DIY

The "user manual" is defined here as a document, either in print or online (although historically produced in small booklet and pamphlet formats, hence the "handbook" alias), which clearly communicates with the widest number of users within a demographic that are engaging with a specific issue or set of issues. A cookbook, for example, is an instructional manual that strives to be useful for anyone who wants to make a particular recipe. If it didn't, it would defeat the purpose. User manuals often focus on the material production of goods, refining and assembling raw materials in order to manufacture any type of object. Equally as prominent are user manuals for using and maintaining equipment, like a toaster or a printer. More recently, user manuals for computer hardware and software have proliferated. The first of the *For Dummies* brand of instructional manuals was published for DOS, the standard disk operating system. Since that publication in 1991, *For Dummies* have gone on to publish instructional manuals on 2,000 individual topics (Graham, 2016). User manuals can also be for building skills and providing services that require no equipment at all, such as mastering the art of Jeet Kune Do, in which Bruce Lee seminally outlines the philosophy and instructions for what is now referred to as mixed martial arts (Lee, 2008). User manuals range dramatically in size and scope, from widely distributed, pocket-sized, illustrated guides like the *Johnson's First Aid Manual*, published in 1929 by the new health-care company called Johnson & Johnson, to narrowly distributed, tome-sized, diagrammatic manuals like the *Guide for Training Nuclear Power Plant Operators*, published by Union Carbide Corporation and the Energy Research and Development Administration in 1966.

Instructional materials are found among the records of ancient antiquity, such as a recently unearthed Egyptian medical papyrus manuscript on face embalming from the era of 3500 BCE (University of Copenhagen, 2021). The Antikythera mechanism, a 2,000-year-old Greek mechanical device that is often referred to as the first computer, has engraved texts that are thought to describe its features and operation, although they have proven notoriously difficult to comprehend by today's researchers (Freeth et al., 2006). These forms of instructional manuals, however, were attached to very specialized objects and divisions of labor, rather than being instructional guides designed to show the masses how they could create on their own. Only later were user manuals in the form of handbooks, the focus of this chapter, produced for the "average" person to use.

The user manual has many subgenres catering to specific professions and disciplines. The sports playbook only operates within the esoteric world of organized sporting, but sports plays do parallel other sectors in the public arena, such as politics and militarization, hence the use of war *games* for training. Field manuals are different from sports playbooks; their emphasis is on being mobile, durable and easy references, in other words, useful in the "outdoors" environment. The environmental conditions where user manuals are used, as well as their subject matter, have led to vast amounts of innovation as people try to provide instruct each other about complex subjects from afar. When Mercedes-Benz produced *the* first car in 1886, they also produced a user manual for it. The Simeone Museum says that this manual pioneered several illustrative and diagrammatic methods for mechanical devices that are still widely used by engineers today. A century after Mercedes-Benz published the first automotive manual, the Haynes company has produced service guides for over 300 models of automobiles. By perfecting illustrative methods such as the "cutout" diagram, accompanied by exhaustive written instructions, the Haynes manuals have greatly contributed to the culture of self-automotive repair, "wrenching" as they call it in the United States. The manuals inspired trends that carried over to other disciplines and were even appreciated purely for their aesthetic qualities, not for their intended use in repairing projects. On the opposite end of the spectrum from the detailed Haynes manuals are the IKEA furniture assembly manuals, which include no text and stick-figure diagrams. I would speculate the IKEA furniture instruction manual is the most widely distributed manual in the history of the world. Their simplistic designs are often mocked in elitist design circles, yet they meet the purpose of expanding the presence of IKEA to as many demographic groups as possible. The benefits become clear after removing all text from the manual. It can now communicate with anyone in the world without requiring the need for translation, thus drastically reducing the size of the manual and avoiding any biases that come with translation, such as choosing languages, dialects and other attributes of syntax.

The examples listed above also fall into the category of being either vertical or horizontal in their publication (and centralized vs. decentralized, to a lesser extent). For example, the Haynes manuals are produced by a third party for other third parties who seek to understand a few hundred different existing products, while an IKEA manual is produced by IKEA for one product that you can only purchase from them. This distinction is important because of its implications for the rise of "maker" or "DIY" culture. Horizontal organizational structures can support many peer-to-peer interactions, while vertical organizational structures can only interact with one level above or below. Peer-to-peer interaction, like we see in open-source systems, have the benefits of freer flow of information, along with building communities of practice.[1] If applied to urban planning, could instructional manuals have the same impact? This idea may not be so far-fetched.

1 A community of practice is a group of people with shared interests who share information and collaborate over significant periods of time using methods of communication.

Historians have recognized one of the earliest instructional manuals as *De architectura*, published as *Ten Books on Architecture*, written by the Roman architect and engineer Marcus Vitruvius Pollio between 30 and 20 BCE (Cartwright, 2020). It contains descriptions on the planning and construction of buildings, military camp sites, and infrastructure such as roads, aqueducts, baths, harbors and more. The book was written as a guide for expanding the empire of Caesar Augustus, not necessarily for the aver.age citizen to deploy on their own volition, but nevertheless it was an instructional guide, and for the development of the city, at that (Vitruvius & Morgan, 1960). The *Ten Books on Architecture* show us the level of value placed on architecture in the development of early civilization. Building homes, streets and other urban infrastructure is the foundation for determining how society will take shape.[2]

The widely acknowledged inception of the user guide in the form of a handbook for the public[3] was *Mechanick Exercises*, published by Joseph Moxon between 1677 and 1683. The book was a technical guide containing step-by-step instructions on blacksmithing, engraving, mapmaking, typography and typesetting, all various aspects of Moxon's practice as a publisher. Beyond providing dry technical information, the book encouraged everyone to take up the maker culture in their own home. Now, many people take the home office or workshop for granted, but at this time it represented a revolution in home design. Moxon's book spurred many Englishmen to convert spare rooms to workshops, learn about machining, and to produce their own wares for use and exchange value. What differentiated Moxon's instructional manual from the guides that had been published before it was its underlying objective to encourage production by anyone, not only by and for the elite (Moxon, 1677–1683). This was the birth of DIY, at least within the sphere of Western culture. At the time, Moxon's book was an implicit attack on the renaissance system of knowledge production and distribution, which was controlled by a network of professional guilds and other exclusive societies. Moxon published information that had previously been exclusively within the domain of the aristocracy and the artisans they chose to support. He also utilized novel language and design practices that made the information accessible to the literate public. By taking knowledge

2 To be clear, not all instructional manuals throughout history have been constructive—such as Bernard Gui's *Practica inquisitionis heretice pravitatis* (Conduct of the Inquisition into heretical wickedness), in which he describes the techniques he used in interrogations as a Spanish inquisitor in the 1300s.

3 Old French (*manuel*): performed by hand, involving physical labor, Classical Latin (*manuālis*): held in the hand, of a size to fill the hand.

out of the domain of elite institutions and making it accessible to the public, *Mechanick Exercises* threatened the monopoly powers of Europe's political and economic institutions (Kaylor, 2021). If you could improve your material surroundings yourself, then who needs the bank or the factory? The discourse that Moxon used also foreshadowed prominent 20th-century trends of "trustless"[4] social collaboration, such as maker spaces[5] and open source[6] information systems, to name a few.

The advent of the public-facing user manual represented several revolutionary concepts. It showed that self-sufficiency in complex matters could be attained through decentralized yet coordinated auto-didacticism. The public was no longer dependent on a "gatekeeper" for education or production of goods; they could teach each other. Access to information through manuals created a multiplier effect in which technology began to develop faster than the guilds and professional societies that had heretofore governed technological development. Access to manuals allowed everyone to be an engineer, to make and adjust designs. That ability to tinker led to new creations, which beget more manuals and novel pedagogies. The instructional manual became increasingly important so that disparate groups could deploy technology without being in direct communication with manufacturers or specialists. The manual is the glue for communities of practice.

From their initial widespread introduction during the Renaissance, private businesses have been quick to adopt the instructional manual format as a means to promote their own objectives. An example is the famed *Merck Manual of Diagnosis and Therapy*, the world's best-selling medical textbook. Initially published as a small reference book in 1899 under the title *Merck's Manual of the Materia Medica* by American drug manufacturer Merck & Co., the manual was compact in size, easy to use and comprehensive. Its novelty, ubiquity and clarity made it a frequently used tool of the industry. It promoted Merck products, but perhaps more significantly, it contributed to the solidification of the foundations of western medicine, such as the definitions of various illnesses, methods for diagnoses and more. The *Merck Manual* is a clear example of how the practice of standardization is used not

4 A trustless system is one in which a third party is not trusted with overriding/centralized control of the system. Open source systems, for example, are written and edited by many users across time and space. No one individual actor can make unilateral decisions that fundamentally change the entire system.

5 A maker space is a community-operated workspace where people with common interests can meet, socialize and collaborate.

6 The term "open source" refers to something people can modify and share because its design is publicly accessible.

only to improve efficiencies in distribution and application of goods, but can also create a conformity of ideology as well. In order to distribute the most drugs, Merck had to align public belief with the system of illness and treatment that Merck products were based on, and it did that through the manual (Winner, 1980). If *Mechanick Exercises* signaled newfound power within the lower classes to engage in material production, then the *Merck Manual* demonstrated the ability of capital to appropriate new mediums of knowledge production and sharing for its own purposes. The *Merck Manual* foreshadowed the further appropriation of DIY discourse by present-day Internet companies, which have repackaged the concept of rent as sharing and collaboration.

Historically, the state has also adopted the method of the user manual to coordinate broad and complex administrative activities. An example is *The Nine Chapters on the Mathematical Art*, which is one of the most important texts in the history of mathematics. It contained arithmetic, algebraic and geometric algorithms for the duties of civil administration such as surveying land, levying taxes, determining wages, measuring grain storage and more. The manual supported autodidactic learning by bureaucrats at a time when managerial systems were not as vast or ingrained as they are now (Shen et al., 2008). State-produced manuals have been used for the implementation of policy, but have gone beyond that to shape the culture of the public sphere. The 1970 New York City Transit Authority (NYCTA) *Graphics Standards Manual* both associated the Helvetica font with New York City and popularized it around the world. The 1975 *NASA Graphics Standards Manual* defined the aesthetic for the fledgling American space program and for the science fiction film genre for a generation (Seemangal, 2015). American government innovation in design is no longer necessarily the case, but manuals are still widely in use by the state. Notably among militaristic organizations, such as *The US Army/Marine Corps Counterinsurgency Field Manual* cowritten by General David H. Petraeus and distributed among US military personnel after America's invasion of Iraq in 2008.

The atlas, a book of maps, represents its own distinct form of instructional manual. From their earliest period, the atlas was the pinnacle of multidisciplinary research and design.[7] The early atlas combined cutting-edge systems of mathematics, and environmental and social research with advanced methods of illustration and printmaking. In 1570, what is thought to be the first global atlas was published by Abraham Ortelius, a Flemish engraver. The atlas, titled *Theatrum Orbis Terrarum* (Theater of the lands of the world),

7 A representation of geography as the "mother discipline."

had both written descriptions and over 150 maps. It pulled from the works of other cartographers of the era and created a standardized system of representation out of them, thus initiating a historical period where maps would become more ubiquitous and usable. Since the publication of Ortelius' *Theatrum Orbis Terrarum*, the atlas has become a critical tool in shaping social interaction, economic production, political power and land development, etc. Atlases, and maps in general, are powerful tools because they project state power, and they engrain sociological distinctions of race and ethnicity, among other things. As tools, they also literally directed human beings where they could and could not go. The cultural impact of the atlas can be seen as widely as the Michelin star system for rating restaurants. Michelin originally produced its atlas, which provides information about dining and lodging options, as a means of selling tires. The Michelin star rating for restaurants now, however, is a cultural tradition that is more coveted and used than the atlas itself (Auguste Escoffier School of Culinary Arts, 2021). The atlas is an excellent example of utilizing geographic information as a form of instructional material. However, the atlas is a tool for navigating human developments, not building them.

The examples above depict the wide range of applications for instructional manuals. The medium of the user guide released knowledge from the confines of elite institutions and spurred localized forms of communication and production. Is it possible then to create a user manual for the city that inverts the process of urbanization as user manuals have done for so many other professions? Could the user manual succeed at empowering the public to challenge state- and corporate-dominated urban planning processes, where conventional models of participation have failed to do so? If all of the pedagogical capabilities of the user manual were applied to the topic of creating a just and sustainable city, what would that user guide look like? What methodologies would it feature? Those questions are explored in the next section, beginning with what the organizational structure of what such a user manual might look like.

The right to repair the city

The "right to repair" is an expansive discourse that covers many different genres of mass-produced consumer goods. It has gained prominence in recent years in regard to the growing integration of consumer electronics in homes throughout America, Europe, Asia and, increasingly, the rest of the world. The complexity of high-tech electronics makes the right to repair a critical issue, given the ability of manufacturers to restrict access to tools and components, or to put in place software barriers, which can hinder independent repair or modification. Preventing repair and modification by consumers can lead to high costs for the consumer and stifles maker culture. Absence of right-to-repair laws also generates massive amounts of environmental pollution since, instead of being reused or upgraded, outdated items are thrown away and new versions are purchased (Šajn, 2022). Concerns about the right to repair first appeared in the United States with the emergence of the automobile industry. While the Ford model of mass production used interchangeable parts across automobile models, making production and repair easy, and keeping costs down, GM developed a new method of continuously "updating" product models as a means of generating demand for parts and service, to which consumers would not have access because of their ever changing designs. By the 1930s, GM's strategy allowed it to overtake Ford as the largest automobile manufacturer; it also started the now widely adopted business model of planned obsolescence.[8] The monopolistic powers of emerging technology companies, such as IBM in the 1960s, made the right to repair a growing issue for consumers since the knowledge and tools needed to repair computers were largely unavailable to the public at that point (Slade, 2007). The COVID-19 pandemic was the impetus for medical right-to-repair legislation proposals in several states, since many health-care providers lacked the ability to service ventilators at the height of hospitalizations (He et al., 2021). The EU has been deliberating on right-to-repair legislation since 2020. Recently, a major step toward implementing widespread regulations was taken when the EU parliament voted to make it a "EU key initiative" (Šajn, 2022). The UK has remained aligned with EU right-to-repair policies via its "eco-design" regulation (Stuart et al., 2021). In June of 2022, New York State passed the first right-to-repair law in the United States (the Digital Fair Repair Act) (Brandom, 2022). Twenty-seven additional states have right-to-repair legislation pending (Proctor, 2021).

When Le Corbusier defined a house (done in the international style) as a "machine for living," he made a strong connection with contemporary urbanism and the mass production of technology. His goal was to turn the city itself into a piece of technology like a wristwatch, where everything operated according to the vision of a singular designer. Unlike the organic

8 Planned obsolescence: a policy of producing consumer goods that rapidly become obsolete and so require replacing, achieved by frequent changes in design, termination of the supply of spare parts and the use of nondurable materials.

and decentralized methods of development that created the historic city of Paris, Corbusier proposed to raze central Paris and replace it with an orthogonal street grid and rectangular skyscrapers (the tower in the park). His plan for Paris, Plan Voisin (named after its benefactor), was never implemented, but the proposal influenced urban development around the world. Cities are now recuperating from the damage that modernist design caused in terms of destroying historic communities and replacing them with technologies like automobile infrastructure and high-rise buildings (Jacobs, 1993). Not only did Le Corbusier inaugurate a new global trend of segregated land uses, he also integrated planned obsolescence into the construction of buildings, in that buildings and their neighborhoods were now viewed as disposable and replaceable, rather than being perpetually repaired, modified and improved. If cities have become a machine for living, should the right to repair not extend to them as well? Ideas like tactical urbanism, DIY urbanism, platform urbanism, hacking the city, etc., have encouraged citizens to repair their cities without official approval. Community groups have painted crosswalks, fixed potholes and made other basic repairs where the state has failed to do so (Tactical Urbanist's Guide, 2022). Groups like the Detroit Technology Project have even installed extensive free public wifi networks in a city where many residents cannot afford the cost of commercially provided Internet (Rodgers, 2017). These types of projects amount to public works done from the ground up. Where community-led public works projects are unimpeded by the state, they have evolved into creating entire self-sufficient districts. The neighborhood of Freetown Christiania in Copenhagen was built by squatters who took over abandoned military barracks. The development began with public works and the creation of a playground (Danish Building and Property Agency, 2022). Now the neighborhood is a frequent destination for visitors seeking to participate in its anticapitalist forms of social engagement. The intense real estate speculation happening in Bushwick, a neighborhood in the New York City borough of Brooklyn, has been fueled for decades by the works of public and private architecture and art created by members of the local community. Squatters and renters used the row houses and loft buildings (former factories) to design otherworldly performance venues, which attracted creators and voyeurs from around the world. But as Bushwick became more famous for its arts culture, the community that created that culture lost the right to repair. Gone are the days of unregulated loft spaces, in which residents came and went freely and interior spaces were constantly made and remade with increasing whimsy. Now, loft spaces are more likely to be regulated by a lease that determines levels of occupancy and what rights the tenant has to design their interior space. Perhaps the most well-known act of community public works is the creation of the hip-hop art form. Hip-hop was born from the necessity to provide music and entertainment in the parks next to the heavily Black and Latino working-class housing communities of New York City in the early 1970s. These housing developments are built in the Corbusian model, a tower surrounded by a "park." In order to provide music and entertainment for gatherings in the grassy areas around the towers, residents would erect two turntables connected with a mixing device, which would allow disc jockeys (DJs) to transition endlessly between various songs. In order to provide electricity to the turntables public parks, DJs would hack public infrastructure (such as a street light) (Morgan, 2009). In this respect, the founders of hip-hop were repairing, modifying and improving the infrastructure of the park by adding music production capabilities, in the process they invented one of the most defining art forms in human history. Through these projects we can see the right to repair is already being exercised by urban dwellers for their public infrastructure, but how powerful would it be if DIY activities were coordinated for the purpose of implementing a people's plan?

In the world of consumer products, when the right to repair exists, manufacturers must utilize designs and make information available that allows for communities of practice to evolve around the maintenance and improvement of those products. Since the city has evolved as a technological product—in a constant state of cost-cutting (austerity) and planned obsolescence—that would put the government in the position of manufacturer. The hardware that is produced by cities of the global north—sewer lids, subway entrances, bridges, street lights, etc.—and the software produced by those cities, their public policies, do not offer the right to repair. They are tightly controlled among state agencies and their private contractors. Right to repair for cities would be governance that enables autonomous actions according to local people's planning, i.e., removing the existing software and hardware barriers.

People's planning

Before talking specifically about creating the user manual, we should talk about why a user manual is needed. What predicates the need for a user manual is the need for people's planning, which is intended here as an insurgent form of urban planning coming from outside of professional institutions, aimed at remaking the city in the image of the working classes.

It operates within the long line of "radical" planning that challenges the authority of state institutions and seeks to develop stronger democratic practices within the planning process (Huq, 2010). Its goal is to displace authoritative plans developed by elite interests and replace them with a community-based vision. This process happens outside of government, but is not forgetful of governments' responsibilities since they created the cities we live in today. In people's planning, community members come together in small and large groups to rethink and reinvent what planning means for them. The new planning framework they create is then filled with research, design and policy ideas of their own.

People's planning emphasizes the constitutive process of building local power, while continuously confronting the question of how locally driven plans might be practically implemented in the face of unresponsive urban political machinery and ever-deepening corporate enclosure of the city. In recent years, political formations such as Cooperation Jackson in Mississippi, the radical municipalism movement in Spain and Take Back the Bronx in New York City have suggested ways in which local neighborhood projects can be woven into broader liberation movements capable of wresting influence over urban territory from both capitalists and unaccountable technocrats, working inside and outside the system at the same time. The practical realities of doing so would give communities the ability to stop their dispossession by having greater community control over the creative process of urban visioning and veto power over local developments. These grassroots resistance movements have exposed the cunning that public and private institutions use to subvert the will of the people. As Shellyne Rodriguez, a leader in people's planning in the South Bronx has written, planning projects in contemporary cities "reveal an unnerving intersection of power that positions real-estate developers, the art world and city government in an alliance to advance gentrification, as a process of systematic repopulation, further into poor and working-class communities" (Rodriguez, 2019). Gentrification is only one example of the historic patterns of land dispossession. At the level of urban planning, that vision for the future is best encapsulated in the Comprehensive Plan of a city or metropolitan area, also known as the strategic plan, master plan, land-use plan, or a variation thereof. Such state-sanctioned forms of planning usually come in lengthy documents that envisage what development and land use will be permitted/constructed for the next 10 to 50 years into the future, ranging from housing and parks to agriculture and much more (Efrussy, 2008). Comp plans have been a fulcrum driving society into the quagmires of climate change, socioeconomic inequality, obesity and other endemic crises. By focusing on the following theoretical principles, people's planning is posited here as their antidote:

- **Grounding in justice:** climate, environmental and social: The climate crisis, like many other social crises afflicting modern cities, is also a crisis of socioeconomic inequality. Therefore, any solutions proposed must be able to address multiple, intersectional problems at once. This requires radical changes in political institutions to enable grassroots community control of land and other resources, as well as a leading role for community stakeholders in any development decisions (Dawson, 2008).
- **Local–global perspective:** In addition to detailed local knowledge, people's planning draws on global examples of radical urban solutions to problems such as the climate crisis. Strive to make your methods of planning replicable and/or interchangeable. That way people's plans can be site specific but use pedagogical and design models that support the growth of a global planning network.
- **Commons:** Human developments everywhere are dependent on a pool of shared resources, whether it's water, air, land or something else. Even though we all share these resources, they can be controlled by a singular entity, be it a business, individual or government, which can lead to exploitation and/or scarcity. Conversely, resources can be controlled democratically by the community members who depend on them—this system is known as the commons. People's planning is a vehicle for learning about the concept of the commons and local methods of application. In the city, the method of the commons can be applied to utilities used every day, such as water and electricity, along with housing, transportation and other key infrastructures.
- **Indigenous practice:** Throughout the colonized world, history typically dates anything about indigenous people as happening in the 19th century or before. There is no sense of continuity from the 19th to the 20th century (Dhillon & Estes, 2019). At this moment we should rethink preservation efforts to include indigenous knowledge that has been systematically excluded over the years. Rather than continue our ignorance of indigenous cultures, we can recognize their lived experience, their leadership in environmental justice activism and their knowledge about natural resources. People's planning seeks to restore the participation and leadership of indigenous people in planning while honoring their lands and rights of self-determination.
- **Security culture:** Governments have become highly proficient at using covert and overt tactics to disrupt community activism. The

tactics developed by the FBI to sabotage racial justice movements in the 1960s (the COINTELPRO counterintelligence program) have only been enhanced since then (Church Committee, 1976). In 2021, it was revealed that governments and private companies easily gained access to software that allowed them to remotely turn on the microphones and cameras on the phones of journalists and political activists, whenever they wanted. The Pegasus spying software, as it's known, is just an example of how profound the threat of intelligence gathering is. Given the amount that governments are investing in advanced computing for the purpose of surveillance, we can only imagine how disruptive the agent provocateurs of the future will be (France 24, 2021). In the context of planning, security culture means being aware of security threats and implementing measures to protect communications, meeting areas and our own physical persons.

- **Solidarity with community mobilizations (social movements):** In contrast to official government planning, people's planning builds on existing grassroots configurations of activists, community leaders and others that demand the right to the city. Social movements are designed, and at least viewed by figures in government, as defined by their stated campaigns and lists of goals, to be placated with negotiations and perhaps some policy changes. People's planning does not view the value of social movements as being limited to specific campaigns, but rather as richly diverse networks of people who can contribute to creating an expansive vision for the future. When organizations or movements have not been limited to one issue or campaign, they have often assumed the role of urban planner or at least community steward. Like the Black Panther Party and its food donation programs, movements have naturally gravitated toward fundamental planning issues. We see that now with the coalitions working to stop fossil fuel pipelines like Dakota Access or Line 3, it's not only professionals or activists working on a singular issue, but wide groups of people who understand the importance of infrastructure planning.
- **Abolitionism:** Planning is a major component of the carceral state. So much so that in rural America there are entire municipalities where prisons, not to mention courts and police officers, are the main feature of the town. In this prison–industrial complex prisons are a major source of employment as well as a source of revenue for funding public infrastructure in addition to being viewed as necessary for keeping the peace and dispensing justice. Without planning none of this would be

possible. Whereas between 1970 and 2005, the number of prisons in the US increased from around 500 to over 1,800, and the rate of imprisoned people grew 700%, people's planning seeks to abolish the institutions responsible for mass incarceration (Misra, 2017). As Ruth Wilson Gilmore, a leader in the study of abolition geography, has described it: "[I]n state and municipal budgets we see the expansion of budgets devoted to mass incarceration, to jails, and to police. We see not only that but, in agencies that are supposed to be working toward other ends—education, health, and so forth—a rise in police functions" (Wilson Gilmore, 2020). It's not regularly discussed but planning and policing have a close connection—the authority of planning agencies is grounded in a legal principle called "police power." Police power is the inherent authority of a government to impose restrictions on anyone in the interest of security, health, safety, morals and welfare. The state retains the power to restrict private and economic interests, including the uses of property (Feldstein, 2007).

- **Self-defense:** In his famous speech "Power Anywhere Where There's People," Fred Hampton recounted how Huey P. Newton and Bobby Seale, leaders of the Black Panther Party (BPP), took matters into their own hands by installing a stop sign at a dangerous intersection in Oakland, California, when the local government failed to do so. Central to the story were BPP members' ability to defend themselves while making the street improvement (Lee, 2019). In an era when there is unfettered surveillance by the state, and local police departments are becoming more weaponized, having the ability to protect oneself, as the BPP advocated, is becoming more important. Maintaining good health, handling oneself in a physical confrontation and otherwise learning how to avoid harmful situations are important aspects of implementing a people's plan.
- **Decolonization:** Whether located in the former imperial metropoles of Europe, settler colonial states in the Americas or Asia, or postcolonial nations still wrestling with the legacy of colonialism and neocolonialism, cities around the planet have been shaped by the histories of exploitation and spatial stratification generated by colonial power. People's planning foregrounds the legacy of settler colonialism and how its present-day manifestations continue to pollute the environment and displace indigenous peoples (Dhillon & Estes, 2019). In the same vein, community planning should be mindful of the impact of proposed solutions in order to ensure that they do not contribute to a new round of neocolonialism and resource extraction in the name of environmentalism.

Table of contents for the user guide

As described earlier, the concept of a user manual adopts several clear principles when it comes to organizational structure and legibility. Instructional manuals depend on disarticulating complex subjects, thus making individual aspects of them more accessible to readers. Breaking down the sum into its parts makes it easier for readers to understand, digest and engage with complex topics. Facilitating proficiency with the parts and the sum of a topic is what allows readers to apply knowledge later for their own individual purposes.

If we are creating a DIY manual for the city, the question then becomes what are the individual elements of the city that will make up the organizational structure of the manual. In other words, what makes a city and, analogously, what constitutes the city planning process? I don't claim to fully answer those questions here, or that they are completely answerable at all, but it's worth speculating what such a set of criteria might be.

This part of this chapter outlines two overarching sections for the structure of the user guide: collaborative process and design implementation. Each section will be detailed below, but the process section includes guidance for social collaboration in the planning process, while implementation includes instructions for constructing and maintaining the built environment.

Before beginning either section, there should be a process of due diligence to uncover existing planning coming from the community. Making the mistake of assuming people's planning has not been done before is a missed opportunity in terms of accounting for existing discourses and building trust with the communities you hope to engage with.

The following research questions and topics can facilitate due diligence:

- Have members of your community collaborated with each other to create a plan that articulates your long-term vision and the research and methods that will be used to get there?
- Do the city planning agencies that govern your community have plans that reflect your community's long-term vision? Is there a need for an alternative plan that can create new outcomes in terms of public health, economics, governance, aesthetics, culture and more?
- Is there an imperative to act before more destruction is inflicted, in part by your planning institutions, on the world's environment, including its human and nonhuman life? Are new systems and institutions required to reverse our path of mass destruction and usher in a future of local and global resilience?

- What was the history of your community before the era of colonial development? Learn what the native plants, animals, topography, watershed and coastlines are. Learn who the communities of native people were that existed there before what has been deemed as "modern history."
- Research current public policies and determine if they reinforce histories of displacement and occupation of land. Consider policies that can correct historical injustices of death and displacement and prevent future ones from occurring.
- Produce a power map diagrams that show the relationships between public institutions, private companies, nongovernmental organizations and other significant stakeholders in your planning area. For each stakeholder, include their name, type of organization, sector and other defining characteristics. Group and arrange them in a way that shows their relationships, including how they overlap and/or oppose each other. Power mapping can help guide strategies, show strengths and weaknesses, and otherwise direct actions (Khawarzad, 2018). Revise the power map throughout the process so it depicts the most up-to-date knowledge.

The collaborative process

Following due diligence, the collaborative process begins. A user manual can provide guidance on forming the collaborative that will manage the planning process. Formation of the collaborative is critical because it will lead the process of rule-making that could shape the full extent of the planning process. Original members of the collaborative are also key players in widening the circle of participants based on recruiting members of their own social networks. These questions are important to keep in mind for community participation in people's planning:

- Who or what are the people? How do they identify themselves—demographically, geographically, culturally, etc.—for the context of this plan?
- Does your planning process have preconceived notions of who comprises "the community" and what their key issues are? Who decides who is included/excluded from the process?
- Who has been wrongly displaced from the land you are occupying? How does your planning process account for their history?

- Are community members themselves able to determine what important planning issues are and how they will be deliberated and agreed upon?
- How can communities develop their own institutions of planning that are autonomous from official structures, even while making claims for accountability and justice upon the latter? How will this planning process organize popular forces at a citywide level, beyond the limited and often corrupt mechanisms of existing electoral politics, and the notoriously toothless format of official public engagement programs?

Research methods

Certain steps in the user manual for cities are to be completed chronologically, whereas research is done persistently throughout the process, serving as a feedback loop for other planning activities such as public outreach, urban design and much more. The research methods below support this style of regenerative planning, but they are not done at the exclusion of other methods. Community planners are encouraged to do their own literature reviews, interviews, environmental data collection and engage in other practices that provide insight into local conditions. The following research methods should be included in people's planning:

- **Participatory action research (PAR):** A fundamental strategy for ensuring community research perspectives and capabilities is integrated into the overall planning process. Planning issues are multilayered, deep and complex. Who is better at understanding the nuances of the city than the people who inhabit it on a daily basis? Technical expertise is emphasized in planning, but often to the exclusion of knowledge coming from community members. For their vast range, professional research methods in planning often fail to account for the informal value of cultures (and subcultures) in cities. There are many forms of communication, personal values, differences in meeting protocols, and a range of other practices that are ignored by the state because of their nonconformity, but are essential for the authentic community representation. PAR is a method of providing a better picture of these on-the-ground conditions.
- **Counter-cartography:** Create maps that communicate people's planning research findings and that reflect the long-term community vision. These maps can be technical documents with precise measurements or expressionist depictions that convey an emotional sentiment about an

area. In contrast to state planning, these maps show the community's interpretation of issues such as borders, important environmental resources and other issues that are typically regarded as the purview of the state or private owners. These maps should speak to the cultures and places where the struggle for justice is most intense and should serve as tools for education and action (Khawarzad, 2021).

Meeting practices

A distinct element of the user manual for just and sustainable cities is its emphasis on facilitating collaborative in-person interactions for urban planning, in other words, meetings. The methodologies used to organize meetings, to facilitate and document discourse, and to do so in ways that empower participants, are key for building long-term participation and a genuine final plan. Some of the methods below can support accomplishing those goals.

- **Self-managed planning hubs:** Popular planning initiatives should establish self-managed planning hubs that provide space for meetings, educational programs, workshops, halls for lectures and film screenings, space for installations including physical models of the city, and more. These hubs should also support broader production and emergency preparedness goals by generating renewable electricity and sustainable agriculture (to the extent possible). Self-management means the collective is responsible for governing and maintaining the physical space, everything from facilitating a process of collective decision-making to switching the lights off at night (Khawarzad, 2018).
- **Forums and assemblies:** Throughout the process there should be workshops, lectures, working meetings, shared meals and other "formal" and "informal" gatherings where critical local knowledge is expressed for integration into the plan. Events can share research findings, collect oral histories, generate quantitative data about environmental issues, connect community members with each other and otherwise serve as critical forms of interaction for planning.

Deliverables

Engaging in a collaborative process accomplishes two goals: building solidarity among community members through interaction, and developing publications that can be distributed among the wider community with the purpose of communicating the planning vision and expanding the collective of participants. The following are some methods for documenting ideas and other commentary that are generated in forums, assemblies, planning hubs and other spaces. Members of the planning collaborative can use these materials to catalyze further participation from the public and as guiding documents for implementing the plan.

- **Comprehensive illustrated vision:** Create a detailed graphic representation of what the shared vision for the future of the "community" is. This illustration helps the community visualize the future, as well as being a method of verifying community input derived during interactive planning sessions (Khawarzad & Fernandez-Muro, 2015).
- **Playbook:** A short document that communicates the key strategies and actions for implementing a people's plan. Creating the playbook is an exercise in answering basic questions about the structure of the planning process and for devising strategies for overcoming institutional challenges.
- **Reference book:** A highly detailed publication that provides instructions for implementing distinct elements of the people's plan. The manual should be published in print and online, and be easily referenced. It serves as the basis of an educational curriculum for groups or individuals who want to contribute to implementing the plan (Khawarzad, 2018).

Design implementation

Creating planning materials at the culmination of the collaborative process is a transitional moment for beginning to focus on designing the implementation of the community plan. Design implementation is also part of the regenerative planning process in that planning, whether being done simultaneously or in the future, will incorporate lessons learned from the implementation of works. This is not to say that aspects of implementation are not discussed in the collaborative process phase, but that there are distinct intellectual and material challenges in the practice of urban development. Coming up with ideas is one thing, but bringing them to fruition is another. The following sections raise the prospects of what the critical topical areas are for a plan. In other words, what subjects, in terms of the material conditions of a city, are the foundation and pillars of a people's plan?

A comprehensive plan for a large city can be intimidating in its vastness, but again, dislocating the process by separating issue areas can make it more accessible. Each issue can be a starting point for group discussions, design exercises and prototype projects for developing the physical environment.

In the historical record, we see "multiple discovery" in how cities evolved throughout history. The understanding that foundational infrastructural and design elements for cities evolved independently of each other throughout various periods in history is some indication that there is an inherent connection between global environmental conditions and how human beings produce urban environments. The orthogonal street grid, for example, is found in the historical record among many disparate cities, disconnected by vast time and space. The city of Ai-Khanoum from the 3rd century BCE is an interesting example. In what is now northern Afghanistan, resting on the banks of the Oxus river, the city is renowned for being easternmost known incursion of Hellenic culture into Asian territories. Ai-Khanoum, or Alexandria on the Oxus, as it was known, has administrative buildings, residences, an arsenal, places of worship, a network of streets, water infrastructure and other functions we would recognize in our contemporary cities (Hoo, 2018). Still in the era of prehistory, around the world in what is now Mexico, the city of Teotihuacan was built along shared philosophies. Although different in its materiality, it shares similar basic tenets with Ai-Khanoum in terms of basic infrastructures, hierarchies of space, land-use types and more. Nowadays we have electricity, the Internet and perhaps other features that we would classify as different from what was in the ancient city, but even new technologies are often integrated into the precedents set by historic forms of urban development. Before Broadway became a street running the length of Manhattan, it was the Wickquasgeck footpath carved by the Lenape people of Mannahatta (Burrows, 2020). New infrastructures (like Internet cables) are placed in old ones (such as water pipes). If we can reverse engineer the key ingredients of what makes a city, then we can write the user manual for it. Here are some potential foundational topics that should be the focus of early planning and development efforts:

- **Economics:** Analyze the socioeconomic demographic makeup of the planning area to develop an understanding of patterns of economic growth and inequality. Consider how economic resources are generated, distributed, and if outcomes of the economic system are fair, sustainable and useful for achieving the goals of the people's plan. Identify alternative economic systems outside of neoliberal capitalism, or whatever system has come to dominate your state planning process, which would support the implementation of your people's plan. As Kali Akuno, member of Cooperation Jackson, writes about the need for community networks to take the reins in building a new economy, "reproducing capitalism, either in its market-oriented or state-dictated forms, will only replicate the inequities and inequalities that have plagued humanity since the dawn of the agricultural revolution" (Akuno & Nangwaya, 2017).

- **Water:** Generate a comprehensive analysis of the watershed within the designated planning area. In your analysis, measure current and projected rates of water usage, and identify where water is being wasted and systems for testing levels of pollution. Urbanization results in multiple stressors to watersheds, such as increased pollutant loads from stormwater runoff, altered stream flows, decreased bank stability and increased water temperatures (University of Michigan, 2022). Work backwards to determine what the original watershed was and how it has been damaged over time. Design water conservation systems, including green infrastructure, methods for reducing consumption at home and work, sustainable agriculture and manufacturing, and more. Implement conservation systems at key locations and then more broadly. Consider the benefits of rewilding the watershed system, examples include daylighting bodies of water and reintroducing native riparian buffers.
- **Waste:** Conduct a comprehensive waste analysis focusing on how to reduce waste by eliminating disposable objects, increasing recycling and reuse, composting and changing aspects of the economy, public policy and cultural behavior that incentivize the creation of waste. Determine the location of waste facilities in your planning area, including landfills, waste transfer stations, incinerators, etc., and analyze their impacts, including disproportionate pollution of communities due to their socioeconomic status. Develop case studies of places that have reduced the size of their waste output. Organize tours of their sustainable waste facilities and document differences in their philosophies of waste (management systems). A town in Japan has become known as a "zero-waste" town. It has no fewer than 45 categories of recycling for household waste, raising the total amount of recycled goods to 80%, when the national average is only 20%. This can only be done with deep levels of participation from the public (Sidjabat & Allm, 2020).
- **Energy:** Conduct a comprehensive survey of the energy generation and consumption of your planning area. Show what the primary sources of energy are, who the largest and smallest consumers are, and other

critical aspects of energy infrastructure. Outline a process for transitioning local energy systems to renewable sources such as solar, tidal, geothermal, wind and more. Show best practices in infrastructure design and prospective relevant technological advancements. Design cooperative organizational structures for constructing and using renewable energy systems. Goals should include reducing consumer costs and increasing local control of infrastructure.

- **Transportation:** Identify areas of improvement for transportation infrastructure, including what should be added, removed or otherwise changed to improve the efficiency (as defined by the community) of all forms of mobility. In your analysis, study traffic flows and their connection to land use to understand how development and economic activity exacerbate transportation challenges. Analyze the economy of the transportation system, including the cost of utilizing necessary modes of transport, along with the economic footprint of the manufacturers and service providers associated with transport systems. Determine the overlap between this economic potential and the people's plan (how can the transportation economy be localized).
- **Flooding:** Map the geographic areas and demographic groups that are most susceptible to flooding from precipitation, extreme rain events, tidal surges and other causes. Show how flooding patterns will be altered as a result of climate change. Study best practices from places that have created protection from flooding. Include practices that insulate cities from flooding with hard infrastructure that channels and segregates water to storage, but also those that use soft infrastructure that permits flooding and allows it to filter back into the watershed more naturally. Prioritize design practices that restore natural landscapes and support diverse plant and animal ecosystems.
- **Housing:** Prioritize planning in neighborhoods that are experiencing the greatest rates of housing insecurity stemming from unaffordability, poor quality, dangerous environmental conditions or other causes. Map these "displacement zones," their physical landscapes and demographics (Khawarzad, 2020). Research examples of housing developments that have architectural designs and governance structures that support environmental sustainability, mental and physical health, and happiness throughout the community.
- **Land:** Land ownership around the world has become extremely concentrated in the outset of the 21st century. In the EU, for example, there are some 12 million farms, and large farms (100 hectares and above) represent 3% of the total number of farms, but control 50% of

all farmed land. In Germany, a total of 1,246,000 land holdings in 1967 shrunk to just 299,100 farms by 2010 (Transnational Institute, 2013). Determine the necessary land resources for implementing the people's plan, including land for housing, open spaces, agricultural use, ecological preservation/regeneration, manufacturing and whatever the plan calls for locally. Conduct an assessment of local land resources including their location, ownership, use, value, etc. Strategize how land resources can be wrested from the control of the gentry and brought within the control of the people. Community land trusts have emerged in recent years as a tactic for decommodifying urban land by removing it from the market and placing it under community control. Identify areas where ancestral indigenous settlements were located and how they were designed. Research historical examples of land restitution where land was taken from indigenous peoples, migrants, ethnic minorities and other oppressed peoples, and which has created systemic inequalities in wealth/access to natural resources.

- **Emergencies:** It has been well-documented that climate change is predicted to increase the frequency and intensity of environmental emergencies. To give an example, one study suggests that the Korean heat wave in the summer of 2013 was ten times more likely due to climate change (Carbonbrief, 2021). The IPCC report from 2021 has called this moment a code red for humanity and says catastrophe is inevitable. Use scenario planning to create an inventory of possible local emergencies resulting from extreme weather, infrastructure failure and other catastrophic events. Show what the potential damages to the community will be in terms of the economy, building stock, natural ecosystems and more. Create emergency prevention and response plans that protect the most vulnerable in the community in times of crisis (and determine who that is). Plans should include improvements to physical infrastructure, communications systems, evacuation transportation, medicine, food and water storage, electrical access and what's needed to meet other common needs. Use self-managed planning hubs to implement emergency plans.
- **Food:** Elucidate the network of food production and distribution that your community is dependent on. Where is your food produced, packaged and how does it get to its final destination? Food accounts for 10 to 30% of a household's carbon footprint, within that production accounts for 68% of food emissions, while transportation accounts for 5% (University of Michigan, 2022). Calculate the environmental impact of the system, and the public health impact of the food being consumed. Revamp the food system based on a local–regional model

that integrates small- to medium-scale agriculture into city planning, and create a food system that operates within a resilient network of native plants, animals and other life. Integrate this system with local economies, transportation infrastructure and housing developments so it works seamlessly and efficiently.
- **Governance:** Make clear what the underlying philosophies and outcomes of your state institutions (if any) are, including how they support and/or serve as obstacles to the objectives of the people's plan. Research alternative models of governance, including those ranging in levels and styles of public participation, and those that support development of the commons.

A critical piece left out of this section is the means by which communities collectively occupy and hold physical space, along with acquiring the labor, knowledge and material resources necessary to develop it according to their plans. This discussion is for another chapter, but suffice it to say, the goal of people's planning is to use the planning process to build the solidarity necessary to undertake the actual city-building process. Perhaps solidarity is built outside of the planning process, but the planning process can support solidarity networks by dovetailing with them in terms of what methods for collaboration are used.

Conclusion

The idea of DIY has always been a threat to the establishment. By providing information and encouragement with a well-designed user interface, entire industries have been undermined by amateurs working out of a spare room in their home. Creative destruction hasn't come from the capitalist class as much as it's come from the autodidacts who are making and sharing information with each other, often for the purposes of self-improvement. Revolutions have been fought through instructional manuals (Powell, 1971). Why can't we use them to build cities, or at least revolutionize the field of community-based urban planning? Of course, the mere creation of a user manual does not mean it's a confrontation of power. Capitalists have proven adept at reorienting DIY activities, and the user manuals they are based on, toward supporting the generation and accumulation of more capital. Uber and Airbnb made everyone a potential delivery driver and innkeeper, but not for the purposes of gaining economic independence, as much as they say that's the case, but for generating more revenue for corporations.

Industrialization in the USA, and now Internet-based companies, have made popular the notion of "self-improvement," where people are told to use their "free time" to take up new hobbies and crafts. The notion of using what would otherwise be time for social reproduction, or simply rest, for "working," can change self-help to exploitation. Remaking the city cannot and should not be a "side project" done in our "free time," but has to be integrated into our daily social practices.

Hopefully this chapter delineates what aspects of DIY urbanism are in line with people's planning vs. what forms of independent action only work to reinforce existing power structures. But, more importantly, this chapter encourages readers to create new systems of planning that are suited for their contexts, and not be completely deferential to state-produced plans that undermine the will of the people.

References

Akuno, K., & Nangwaya, A. (Eds.). (2017). *Jackson rising: The struggle for economic democracy and Black self-determination in Jackson, Mississippi*. Daraja Press.

Auguste Escoffier School of Culinary Arts. (2021). How restaurants get Michelin stars: A brief history of the *Michelin Guide*. https://www.escoffier.edu/blog/world-food-drink/a-brief-history-of-the-michelin-guide/

Brandom, R. (2022, June 3). New York state passes first-ever "right to repair" law for electronics, *The Verge*. https://www.theverge.com/2022/6/3/23153504/right-to-repair-new-york-state-law-ifixit-repairability-diy

Burrows. E. G. (2000). *Gotham: A history of New York City to 1898*. Oxford University Press.

Carbonbrief. (2021). Mapped: How climate change affects extreme weather around the world. https://www.carbonbrief.org/mapped-how-climate-change-affects-extreme-weather-around-the-world/

Cartwright, M. (2020). Vitruvius. In *The world history encyclopedia*. https://www.worldhistory.org/Vitruvius/

Church Committee. (1976). The Select Committee to Study Governmental Operations with Respect to Intelligence Activities, Foreign and Military Intelligence. Church Committee report, no. 94–755, 94th Cong., 2d Sess. United States Congress.

Climate Action Lab. (2019). *A people's climate plan for New York City?* The Graduate Center, City University of New York. https://occupyclimatechange.net/wp-content/uploads/2022/03/ClimatePlan_FINAL1-1.pdf

Danish Building and Property Agency. (2022). History of Christiana area. https://en.bygst.dk/

Dawson, A. (2017). *Extreme cities: The peril and promise of urban life in the age of climate change*. Verso.

Dhillon, J., & Estes, N. (Eds.). (2019). *Standing with Standing Rock: Voices from the #NoDAPL movement*. University of Minnesota Press.

Efrussy, A. M. (2008). *Introduction to the comprehensive plan: A guide for urban planning*. Texas Educational Foundation.

Feldstein, L. M. (2007). *General plans and zoning: A toolkit for building healthy, vibrant communities*. California Department of Health Care Services.

France 24. (2021, July 18). Private Israeli malware "Pegasus" used to spy on journalists, activists and politicians. https://www.france24.com/en/technology/20210718-private-israeli-malware-used-to-spy-on-journalists-activists-and-politicians

Freeth, T., Bitsakis, Y., Moussas, X., et al. (2006). Decoding the ancient Greek astronomical calculator known as the Antikythera mechanism. *Nature, 444*, 587–591. https://doi.org/10.1038/nature05357

Graham, R. (2016, April 4). Dummies for dummies, *Slate*. https://slate.com/culture/2016/04/the-history-and-delights-of-the-for-dummies-how-to-books.html

He, S., Lai, D., & Lee, J. (2021). The medical right to repair: The right to save lives. *Lancet*, 397(10281), 1260–1261. https://doi.org/10.1016/S0140-6736(21)00445-1

Hoo, M. (2018). Ai Khanum in the face of Eurasian globalisation. *Ancient West & East, 17*, 161–186. https://doi.org/10.2143/AWE.17.0.3284893

Huq, E. (2020). Seeing the insurgent in transformative planning practices. *Planning Theory, 19*(4), 371–391. https://doi.org/10.1177/1473095219901

Jacobs, J. (1993). *The death and life of great american cities*. Vintage Books.

Johnson & Johnson Ltd. (2018). *Johnson's first aid manual: The classic 1929 edition*. Clydesdale.

Kaylor, B. (2021, May 3). How to: On the unlikely political and cultural power of the DIY manual. *Literary Hub*. https://lithub.com/on-the-unlikely-political-and-cultural-power-of-the-diy-manual/

Khawarzad, A. (2020). *New York City climate change displacement map*. Rosa Luxemburg Stiftung. https://rosalux.nyc/climate-change-displacement-map-of-new-york-city/

Khawarzad A. (Ed.). (2018). *The Upper Manhatta(n) Project*. AKK Press.

Khawarzad, A., & Fernandez-Muro, M. (2015). *Northern Manhattan climate action plan*. https://www.weact.org/campaigns/nmca/

Lee, B. (2008). Tao of Jeet Kune Do. Black Belt Communications. (Original edition 1975).

Lee, J. (2008, January 15). Tenants might buy the birthplace of hip-hop, *New York Times*.

Lee, W. (2019, December 3). In 1969, charismatic Black Panthers leader Fred Hampton was killed in a hail of gunfire: 50 years later, the fight against police brutality continues. *Chicago Tribune*. https://www.chicagotribune.com/news/ct-black-panthers-raid-fred-hampton-50-years-20191203-kbzgztrvtfh7tp7x4ggtvhncpm-story.html

Merck & Co. (1899). *Merck's manual of the Materia Medica.*

Misra, T. (2017, May 2). The economics of prison boomtowns, *Bloomberg*. https://www.bloomberg.com/news/articles/2017-05-02/inside-the-prison-towns-of-the-rural-south

Morgan, L. (2009, May 19). 35 years of hip hop: It began with hacking a lamp post. *Wired*. https://www.wired.com/2009/05/35-years-of-hip-hop-it-began-with-hacking-a-lamp-post/

Moxon, J. (1677–1683). *Mechanick exercises, or, The doctrine of handy-works*. J. Moxon.

NASA. (1975). *NASA graphics standards manual*. National Aeronautics and Space Administration.

NYCTA. (1970). *Graphics standards manual*. New York City Transit Authority.

Petraeus, D. H., & Amos, J. F. (2007). *The US Army/Marine Corps counterinsurgency field manual.* US Army Field Manual no. 3-24. Marine Corps Warfighting Publication no. 3-33.5. University of Chicago Press.

Powell, W. (1971). *The anarchist cookbook*. L. Stuart.

Proctor, N. (2021, March 10). Half of US states looking to give Americans the right to repair. *PIRG*. https://pirg.org/articles/half-of-u-s-states-looking-to-give-americans-the-right-to-repair/

Rodgers, K. (2017, November 16). Ignored by big telecom, Detroit's marginalized communities are building their own Internet. *Vice*. https://www.vice.com/en/article/kz3xyz/detroit-mesh-network

Rodriguez, S. (2019, December 12). How the Bronx was branded. *New Inquiry*. https://thenewinquiry.com/how-the-bronx-was-branded/

Šajn, N. (2022). European Parliament right to repair briefing. European Parliamentary Research Service. https://www.europarl.europa.eu/RegData/etudes/BRIE/2022/698869/EPRS_BRI(2022)698869_EN.pdf

Seemangal, R. (2015, September 28). NASA is the unlikeliest "design firm" in human history, *The Observer*. https://observer.com/2015/09/nasa-is-the-unlikeliest-design-firm-in-human-history/2015

Shen, K., Crossley, J. N., Lun, A. W. C., & Liu, H. (2008). *The nine chapters on the mathematical art: Companion and commentary*. Oxford University Press.

Sidjabat, F. M., & Ilmi, A. (2020). Kamikatsu Japan's ecovillage conceptual framework for environmental management. *Journal of Physics Conference Series, 1625*(1), 012063. https://doi.org/10.1088/1742-6596/1625/1/012063

Slade, G. (2007). *Made to break: Technology and obsolescence in America*. Harvard University Press.

Stuart, G., Hemmings, J., Brown, H., MacLeod, R., & Haque, A. (2021, August 5). EU and United Kingdom: UK implements new "right to repair" rules for household products to align with EU ecodesign requirements. *Global Compliance News*. https://www.globalcompliancenews.com/2021/08/05/eu-and-united-kingdom-uk-implements-new-right-to-repair-rules-for-household-products-to-align-with-eu-ecodesign-requirements210721/

Tactical Urbanist's Guide. (2022). About tactical urbanism. http://tacticalurbanismguide.com/about/

Transnational Institute. (2013). Hands Off the Land: Take action against land grabbing. https://www.tni.org/en/collection/hands-off-the-land

Union Carbide Corporation and the Energy Research and Development Administration. (1966). *Guide for training nuclear power plant operators, Springfield (VA)*. National Technical Information Service and US Department of Commerce.

University of Copenhagen. (2021, February 26). Ancient Egyptian manual reveals new details about mummification. https://humanities.ku.dk/news/2021/ancient-egyptian-manual-reveals-new-details-about-mummification/

University of Michigan (2022, September). Carbon Footprint Factsheet. Pub. No. CSS09-05. https://css.umich.edu/sites/default/files/2022-09/Carbon%20Footprint_CSS09-05.pdf

Wilson Gilmore, R. (2020, June 10). Ruth Wilson Gilmore makes the case for abolition. *The Intercept*. https://theintercept.com/2020/06/10/ruth-wilson-gilmore-makes-the-case-for-abolition/

Winner, L. (1980). Do artifacts have politics? *Daedalus, 109*(1), 121–136.

Vitruvius, P., & Morgan, M. H. (1960). *Vitruvius: The ten books on architecture*. Dover Publications.

About the author

Aurash Khawarzad is an artist, educator and urban planner. His work uses a combination of research, creative practice and multidisciplinary collaboration as a means of visioning communities. He is currently a PhD student in geography at the Graduate Center, City University of New York. He is also the author and publisher of *The Upper Manhatta(n) Project*, an international model for climate change planning.

Index

For Product Safety Concerns and Information please contact our EU representative GPSR@taylorandfrancis.com Taylor & Francis Verlag GmbH, Kaufingerstraße 24, 80331 München, Germany

Batch number: 10725973

Printed by Printforce, the Netherlands